Restructuring the U.S. Steel Industry

Restructuring the U.S. Steel Industry

Semi-Finished Steel Imports,
International Integration,
and U.S. Adaptation

Jose Guilherme de Heraclito Lima

Westview Press
BOULDER • SAN FRANCISCO • OXFORD

This Westview softcover edition is printed on acid-free paper and bound in library-quality, coated covers that carry the highest rating of the National Association of State Textbook Administrators, in consultation with the Association of American Publishers and the Book Manufacturers' Institute.

Tables 3.9, 3.10, 3.14, 5.1, 5.2, 5.4, 5.5, 5.7, 5.8, 5.9, 6.1, A.3, A.10, A.12, and B.1 are reprinted by permission of the American Iron and Steel Institute. Tables 6.5, 6.7, and 6.10 are reprinted by permission of the Brookings Institution. Tables 3.1, 3.5, 3.6, 3.7, 3.8, 3.16, 3.17, 3.18, 3.19, 3.21, and 3.24 are reprinted by permission of the International Iron and Steel Institute. Table 3.28 is reprinted by permission of the Massachusetts Institute of Technology. Table 5.14 is reprinted by permission of 33 Metal Producing. Tables 3.20, 3.21, 3.23, 3.25, 3.26, and 3.27 are reprinted by permission of the United Kingdom Iron and Steel Statistics Bureau. Table 3.13 is reprinted by permission of the WEFA Group. Tables 3.2, 3.3, 3.4, 3.11, 3.15, 4.1, 4.2, 4.3, 4.4, 4.7, 5.9, 6.2, 6.9, 6.13, 6.22, 7.1, 7.2, A.1, A.2, A.5, A.6, A.7, A.8, A.13, B.1, and B.2 are reprinted by permission of World Steel Dynamics.

Table A.4 is reprinted by permission of *The Wall Street Journal,* copyright 1985 Dow Jones & Company, Inc. All Rights Reserved Worldwide.

All rights reserved. No part of this publication may be reproduced or transmitted in any form or by any means, electronic or mechanical, including photocopy, recording, or any information storage and retrieval system, without permission in writing from the publisher.

Copyright © 1991 by Westview Press, Inc.

Published in 1991 in the United States of America by Westview Press, Inc., 5500 Central Avenue, Boulder, Colorado 80301, and in the United Kingdom by Westview Press, 36 Lonsdale Road, Summertown, Oxford OX2 7EW

Library of Congress Cataloging-in-Publication Data
A CIP catalog record for this book is available from the Library of Congress.
ISBN 0-8133-8043-X

Printed and bound in the United States of America

∞ The paper used in this publication meets the requirements of the American National Standard for Permanence of Paper for Printed Library Materials Z39.48-1984.

Contents

List of Tables ix
Foreword, Richard L. Gordon xv
Acknowledgments xix

1 Introduction 1

    An Overview of the Book, 3

2 Existing and Emerging Steelmaking Technologies and Their Market Implications 5

    An Overview of Steel Production, 5
    Coking Coal and Coke Ovens, 9
    Alternative Iron-Ore Reduction Processes, 10
    Steelmaking Technology, 11
    Continuous Casting and Flat-Rolled Finishing Technology, 13
    Summary and Conclusions, 15

3 World and United States Markets for Steel and Steelmaking Raw Materials 16

    Consumption and Production Trends and World Steel Excess Capacity, 16
    International Steel Trade, 20
    The United States Steel Market, 28
    Effects on the Production and Trade of Iron Ore, Scrap, and Semi-Finished Products, 37
    Summary and Conclusions, 58

4 Adaption of the U.S. Steel Industry to Market Trends 61

    Raw Material Producing Capacity Reduction, 61
    Steel Plant Closings, 62

The Regional Distribution of the Crude Steel
  Integrated Capacity, 67
Closing of Integrated Mill Finishing Capacity and Mini-Mill
  Capacity Development, 67
U.S. Flat-Rolled Capacity, 68
Restructuring and the Competitiveness of Existing Plants, 68
Summary and Conclusions, 72

5 The Emergence of Imported Slabs in the U.S. Market and Its Implications for Integrated Steel Producers — 73

The Flat-Rolled and Semi-Finished Steel Markets and
  Their Regional Differences, 73
The Deficiency of Steelmaking Capacity of the U.S. Flat-Rolled
  Producers and Its Impact on Market Structure, 85
Other Factors Contributing to Slab Imports, 91
Summary and Conclusions, 95

6 Steel Production Costs and the Prospects for Slab-Steel Imports — 96

Principles of Supply Analysis, 97
An Overview of Comparative Input Supply Problems of Steel
  Producers in the United States and Abroad, 100
An Overview of Steel Comparative Economics, 104
The Competitiveness of Purchasing Slab or DRI by U.S.
  Flat-Rolled Carbon Steel Producers, 114
The Profitability of an Integrated Producer and a
  Non-Integrated Slab Purchaser, 122
The Competitiveness of Importing Slab, 129
Summary and Conclusions, 132

7 Protectionism in Steel — 134

The Economics of Protectionism, 134
U.S. Trade Barriers and the Impact of VER, 135
Slab Import Restriction and Its Effects on Domestic
  Steel Producers, 140
Summary and Conclusions, 146

8 Summary and Conclusions — 147

Appendixes

A The Financial Situation of U.S. Integrated Producers and the Implications for Investment — 151

B	**Closures of Steel Capacity**	**166**
C	**Capacity Balance Equations**	**171**
D	**Cost Estimation Procedure**	**174**

Bibliography 177

Tables

3.1	Apparent Steel Consumption and Crude Steel Production by Region--1973 to 1986	18
3.2	World Gross Capacity and Gross Operating Rate by Region--1973 to 1986	20
3.3	New Greenfield Steel Plants Planned for 1985-95	21
3.4	Noncommunist World Effective Capacity and Effective Operating Rate by Region--1973 to 1986	22
3.5	Exports, Imports, and Net Exports of Semi-Finished and Finished Steel by Region--1973 to 1986	23
3.6	Main Net Exporters and Net Importers of Semi-Finished and Finished Steel in 1986	24
3.7	Apparent Consumption, Crude Steel Production, and International Trade in Semi-Finished and Finished Steel by Brazil, South Korea, South Africa, and Rumania--1973 to 1986	25
3.8	Apparent Steel Consumption, Crude Steel Production, and International Trade of Semi-Finished and Finished Steel by Regions of the Noncommunist World--1977 to 1986	26
3.9	Net U.S. Steel Mill Products Shipments to Main Sectors--1973 to 1987	30
3.10	U.S. Steel Industry Capacity, Crude Steel Production, International Trade, Apparent Steel Consumption, Net Shipments, and Yield--1956 to 1987	34
3.11	U.S. Major Mill Actual and "Breakeven" Operating Rates by Quarters--1981 to 1986	35
3.12	U.S. Net Manufacturers' Shipments, Exports, Imports, and Apparent Consumption of Steel by Steel Grade and U.S. Export and Import Prices--1973 to 1987	36
3.13	Major U.S. Steel Producers and Their Product Lines in 1981 and 1985	37
3.14	Net Shipments of U.S. Steel Mills by Major Products Categories, All Grades--1973 to 1987	38

3.15	U.S. Net Trade in Steel-Containing Goods	39
3.16	Apparent Consumption, Production, and International Trade of Iron Ore by Regions of the World--1977 to 1986	40
3.17	Apparent Consumption, Production, and International Trade of Pig Iron by Regions of the World--1977 to 1986	42
3.18	Direct Reduction Production by Regions of the World--1977 to 1986	44
3.19	Consumption, Apparent Production, and International Trade of Scrap by Regions of the World--1977 to 1986	45
3.20	International Trade of Semi-Finished Steel by Regions of the Noncommunist World--1977 to 1985	50
3.21	Share of Semi-Finished Steel Imports for Selected Countries--1977 to 1985	52
3.22	U.S. Carbon Steel Slab Import by Country of Origin--1984 to July 1988	53
3.23	Semi-Finished Steel Imports as Percent of Crude Steel Production in Selected Countries--1977 to 1985	54
3.24	Share of World Imports of Semi-Finished Steel for Selected Countries--1977 to 1985	55
3.25	Main Exporters of Semi-Finished Steel in 1987	56
3.26	Imports Semi-Finished Product Mix by Regions of the World--1977 to 1985	57
3.27	Imports of Hot-Rolled Coil for Re-Rolling by Regions of the World--1979 to 1985	59
3.28	Semi-Finished Steel Plants Planned or Considered for 1980	60
4.1	U.S. Coke Oven Capacity, Production, Operating Rate, and U.S. Coke Imports--1973 to 1986	63
4.2	U.S. Blast Furnace Capacity, Production, Operating Rate, Coke Rate, and Number--1973 to 1986	64
4.3	U.S. Steelmaking and Continuous Casting Capacities, Production, and Operating Rates--1973 to 1986	65
4.4	U.S. Crude Steel Capacity Ownership by Type of Firm--1977 to 1986	66
4.5	Share of the Six Largest Integrated Producers in the U.S. Steel Shipments--1950 to 1986	67
4.6	U.S. Mini-Mills Production, Capacity and Utilization of Steelmaking for Selected Products of Carbon and Alloy Steel from 1984 to 1988	69
4.7	U.S. Mini-Mills Producers' Shipments by Selected Products of Carbon and Alloy Steel from 1984 to 1988	70
4.8	Share of Largest Producers in U.S. Flat-Rolled Capacity--1977 to 1986	70
4.9	Aggregate U.S. Integrated Carbon Steel Industry Facility Rankings in 1988	71
5.1	U.S. Shipments, International Trade, and Consumption of Flat-Rolled Products by Main Product Categories--1979 to 1987	75

Tables   xi

#	Title	Page
5.2	U.S. Domestic Flat-Rolled Product Supply by Grade--1979 to 1987	76
5.3	U.S. Shipments, Imports and Prices of Carbon Flat-Rolled Carbon Steel by Type--1979 to 1987	77
5.4	Role of Leading Suppliers in U.S. Flat-Rolled Product Imports--1979 to 1987	79
5.5	U.S. Shipments, International Trade, and Consumption of Semi-Finished Steel--1979 to 1987	80
5.6	U.S. Shipments, Imports and Prices of Semi-Finished Steel by Grade--1979 to 1987	81
5.7	U.S. Semi-Finished Steel Imports by Country of Origin--1979 to 1987	83
5.8	U.S. Imports of Selected Steel Products by Region--1979 to 1987	85
5.9	Estimated Ratio of Import to Domestic Production of Flat-Rolled and Semi-Finished Products by Region and Product Type--1979 to 1987	86
5.10	Excess of Steelmaking Capacity over U.S. Flat-Rolled Product Finishing Capacity for U.S. Flat-Rolled Producers at Different Operating Rates--1979 to 1986	88
5.11	Excess of U.S. Steelmaking Capacity over U.S. Flat-Rolled Product Finishing Capacity for U.S. Flat-Rolled Producers at Different Operating Rates--1979 to 1986	89
5.12	Excess of U.S. Coking Capacity over U.S. Blast Furnace Capacity at Different Operating Rates--1979 to 1985	93
5.13	Excess Capacity of U.S. Blast Furnace Capacity over U.S. Steelmaking Capacity Both at 90 Percent Operating Rate and Different Scrap Charging--1979 to 1985	93
5.14	Estimated U.S. Coke Oven and Blast Furnace Capacity Requiring Relining	94
6.1	Number of Employees and Total Employment Costs per Hour Worked in the U.S. Steel Industry--1973 to 1987	103
6.2	Typical Contract and Spot Ocean Freight Rates and Customs and Other Costs Associated with Delivering Steel Products to the United States--1973 to 1986	105
6.3	Nominal Exchange Rates for the Main Countries Exporting to the U.S. Market--1973 to 1987	106
6.4	Real Exchange Rates for the Main Countries Exporting to the U.S. Market--1973 to 1987	107
6.5	Cost of Producing Wire Rod at Representative U.S. Mini-Mills and Integrated Producers in the U.S. and Japan in 1985	108
6.6	U.S. Import Penetration and Ratio of Imported and Domestic Prices for Selected Products--1977 to 1987	109
6.7	Cost of Producing Cold-Rolled Sheet at an Efficient Integrated Plant in the United States and Four Selected Countries in 1985	110

6.8	Cost of Producing Cold-Rolled Sheet at an Efficient Integrated Plant in the United States and Four Selected Countries in 1986	111
6.9	Cost of Producing Cold-Rolled Sheet at an Efficient Integrated Plant in the United States and Four Selected Countries in 1987	112
6.10	Impact of Dollar Devaluation Against the Mark and Yen on the 1985 Cost of Producing Cold-Rolled Sheet	113
6.11	U.S. Flat-Rolled Producer 1987 Capacity Assessment	114
6.12	1987 Costs of Production, Prices, and Profit Margin for U.S. Class 1 and 3 Integrated Producers	115
6.13	Home Country Composite Steel Prices of the United States, West Germany, and Japan in Dollar and Home Countries' Currency--1980 to 1987	116
6.14	Cost of Producing Slab at Efficient Integrated Producers for U.S. Inland and Coastal Regions and Four Selected Countries in 1985	117
6.15	Cost of Producing Slab at Efficient Integrated Producers for U.S. Inland and Coastal Regions and Four Selected Countries in 1988	118
6.16	Transportation Cost from Exporting Regions and Domestic Producers to Selected Regions in the United States in 1988-89	120
6.17	Estimated Slab Price Delivered at the Plant Location in 1985 and 1988	121
6.18	Estimated Total Composite Production Cost at a U.S. Efficient and Inefficient Non-Integrated Producer Purchasing Slab	122
6.19	Profit Margin at a U.S. Efficient and Inefficient Non-Integrated Producer Purchasing Slab	123
6.20	Estimated Total Composite Production Costs at a U.S. Integrated Producer by Region, Iron-Making Technology Used, and Type of Plant	124
6.21	Estimated Profit Margin at a U.S. Integrated Producer by Region, Iron-Making Technology Used and Type of Plant	125
6.22	Estimates of Investment Options of a U.S. Flat-Rolled Producer and Its Behavior	126
6.23	Net Present Value of a Green-Field Plant Producing Flat-Rolled Products by Type of Plant for Selected U.S. Regions Producer Behavior at 6 Percent and 8 Percent Discount Rates	127
6.24	Net Present Value of Buying an Existing Integrated Plant with Efficient and Inefficient Finishing-End for Selected U.S. Regions and Producer Behavior at 6 Percent and 8 Percent Discount Rate	128

Tables   xiii

6.25	Net Present Value of Closing the Hot-End at a Plant with Efficient and Inefficient Finishing-End for Selected U.S. Regions and Producer Behavior at 6 Percent and 8 Percent Discount Rate	130
6.26	Net Present Value of Replacing Hot-End Facilities at an Existing Integrated Plant with an Efficient Finishing-End for Selected U.S. Regions and Producer Behavior at 6 Percent and 8 Percent Discount Rate	131
6.27	U.S. International Competitiveness in the Flat-Rolled Market of Integrated Producers, Non-Integrated Producers Purchasing Slabs, and Importing Slab, Compared to Imports from Brazil and South Korea in 1988	132
7.1	U.S. Steel Trade Agreements for Finished Steel in 1987	139
7.2	U.S. Steel Trade Arrangements for Semi-Finished Steel in 1987	141
7.3	Short Supply Requests for Slab by U.S. Steel Mills in 1987 and 1988	143
A.1	Percent of Return on Sales for the Six Largest U.S. Integrated Steel Producers and Nucor--1973 to 1986	152
A.2	Net Income and Operating Rate of the Six U.S. Largest Integrated Steel Producers and Nucor--1981 to 1987	153
A.3	Indicators of U.S. Steel Industry Return on Investment and Assets and the Capital Intensity--1979 to 1987	154
A.4	Market Value of Common Stocks Compared to the Share Asset Value of the Six U.S. Largest Integrated Steel Producers and Nucor in 1984	156
A.5	Earnings per Share and Stock Prices for the U.S. Steel Industry and the U.S. Standard & Poor's 400--1973 to 1986	157
A.6	Earnings per Share for the Six Largest U.S. Integrated Steel Producers and Nucor--1973 to 1986	158
A.7	Stock Prices for the Six Largest U.S. Integrated Steel Producers and Nucor--1973 to 1986	159
A.8	Price Earnings Ratio for the Six Largest U.S. Integrated Steel Producers and Nucor--1973 to 1986	160
A.9	Moody's Bond Rating of the Six Largest U.S. Integrated Steel Producers--1980 to 1987	161
A.10	Indicators of the Financial Condition of the U.S. Steel Industry--1979 to 1987	162
A.11	Debt as a Percent of Equity for the Six Largest U.S. Integrated Steel Producers--1984 to 1986	163
A.12	Total U.S. Steel Industry Capital Expenditures and Outlays for Pollution Control--1960 to 1986	164
A.13	Cost Difference Between Reorganized and Other U.S. Integrated Producers in 1987	165

B.1	Regional Distribution of the U.S. Crude Steel Capacity--1977 to 1986	168
B.2	Regional Distribution of the U.S. Flat-Rolled Capacity--1977 to 1986	169

Foreword

Dr. Lima's book reviews developments in the world steel market and the prospects for further changes. Dr. Lima, when preparing this study, was an employee of Siberbras--Brazil's government-owned steel company--and was particularly interested in changes that would benefit the company. His focus is on concerns that production of high-quality raw steel for flat-rolling is more expensive in the U.S. than in some foreign countries.

He evaluates the possibilities of countries such as Brazil, Venezuela, and South Korea becoming the lowest-cost suppliers of U.S. mills for slabs needed to produce finished rolled products. His conclusion is that the existing and planned capacity in these countries is large enough and low enough in cost economically to supply in the range of five to ten million (metric) tons of additional slab to U.S. mills.

This conclusion is supported by an extensive review of world and U.S. steel industry trends, an overview of technical trends in steel making, and the most accurate available data on producing capacity and costs in steel making. Before commenting on the issues raised by Dr. Lima's analysis, I'd like to review some of the basic information.[1]

World steel output and consumption peaked at about 745 million tons in 1979, declined through 1982, and then rose to around 715 million tons in 1986. (1988 output was 779 million.) These developments are the net effect of declines in industrialized countries and increases in the CPE and developing noncommunist countries. Consumption and production in the industrialized countries peaked around 1973; growth tended to prevail from 1973 to 1979 in the rest of the world. Thus, 1979 was when the decline in the industrialized countries began to exceed rises elsewhere.

Since 1986, capacity in industrialized countries was at roughly the 1973 level, capacity utilization rates fell. A smaller decline in utilization rates prevailed for CPE countries; utilization rates rose for developing countries.

World trade also expanded over the 1973-86 period--from around 110 million to 156 million tons. The three main elements of the change were an 8 million-ton rise of U.S. net imports, a 7 million-ton fall in net imports by developing countries, and a 15 million-ton increase in net imports by the CPE countries (predominantly China and the U.S.S.R.). Brazil, South Korea, South Africa, and Rumania were the main sources of expanded

---

[1] The review comes from the summary of his chapter 3. See the report for details on the definitions of the country groups used.

exports. Another contribution came from iron-rich, oil-producing countries, mainly Venezuela and Mexico.

Examination of U.S. mill shipment data shows that all direct purchasers of steel reduced purchases from U.S. mills. A move to increased purchasing through service centers in the 1980s brought 1987 shipments to service centers to 18 million tons, compared to 18.5 million tons in 1973.

Smaller scaled operations termed mini-mills have arisen in the United States and have begun to account for the majority of U.S. industry shipments of non-flat-rolled products such as rods, bars, and structural shapes. These mini-mills use electric-arc furnaces to make steel from scrap. Flat-rolled products still are produced predominantly by integrated mills; flat-rolled products, in turn, constitute the bulk of integrated-mill output.

The decline in steel production from 1979 levels left 1986 world production of iron ore and pig iron and consumption of iron ore below 1979 levels. The much smaller sector producing direct reduced iron expanded from 6.9 million tons in 1979 to 12.6 million tons in 1986.

The regional composition of iron ore, pig iron, and scrap supply also changed. The main trends in iron ore were increases in the CPE and iron ore-rich developing countries and declines in the U.S., the EEC, and the iron-rich industrialized countries. Brazil was the main source of iron-ore output rises. Since pig-iron trade changed little, pig-iron production trends tended to follow steel production.

Scrap generation comes predominantly from industrialized noncommunist and CPE countries. Supplies declined in the industrialized noncommunist countries and remained steady in the CPE countries.

International trade of iron ore, scrap, semi-finished steel, and finished steel were higher in 1986 than in 1977. However, iron-ore exports peaked at 395 million tons in 1979. The 1986 level of 366 million tons was 7.5 percent below the 1979 level of 396 million tons. The main regional trends were a steady rise since 1973 of exports from iron-rich developing countries and a decline starting after 1979 that brought 1986 exports of iron-rich developed countries below their 1977 levels.

Scrap and semi-finished steel had small international markets with the largest percentage increases. Both markets had 1986 levels above 1979. The scrap market increased by 67 percent to 30 million tons in 1986. The U.S. is by far the largest net exporter of scrap and had by far the largest change in net exports.

The semi-finished market grew by 66 percent to 13.5 million tons. Iron-ore trade increased by 3 percent and finished steel trade grew by 19 percent. The main net sources are the EEC and the iron-rich developing countries, particularly Brazil. Dr. Lima's synthesis of capacity and cost data lead him to conclude that further growth of similar magnitude is possible for semi-finished steel.

As Dr. Lima is aware, this is only one of several alterations in steelmaking technology that might further change industry patterns. Development of both iron-making and flat-rolling technologies economical at small scales might lead to the rise of small mills in the flat-rolled product area. Conversely, new technologies might reduce costs at the large-scale mills that presently produce flat-rolled products.

Foreword

Two issues that lie beyond the scope of this book are the original wisdom of investments in steel mills in Brazil and other countries and the ability of different producers to attain the quality standards demanded by U.S. mills and their customers. The first issue is one on which the U.S. steel industry bases its demands for protection.

Thus, in 1988, a report was conducted by the law firm Dewey, Ballentine, Bushby, Palmer & Wood for the six largest U.S. steel companies (Armco, Bethlehem, Inland, LTV, National, and USX). The report reviews steel industry capacity expansion throughout the world.[2] Many countries are faulted for ill-conceived expansions and subsidies for their continued existence. The report concludes that this justifies protection. This view is at least as problematic as the free trade stance taken by Dr. Lima. The Dewey-Ballentine report neglects the basic economic concept of sunk costs. The clearly correct parts of the Dewey-Ballentine charges demonstrate a deplorable state of policy making around the world.

However, most of these mistakes involve creation of long-lived facilities. The world must go on to maximize economic efficiency given these plants.

If these plants can sell at prices sufficient to recover future costs including both cash costs of production and new maintenance and investment outlays, the plants should be kept going. What is unwise is provision of new subsidies to maintain plants that cannot recover their future costs.

Even here, severe limits exist on what is desirable and practical for the U.S. to do unilaterally. The subsidies are a gift of the rest of the world to the U.S., and the U.S. economy would be worse off if it restricted receipt of these gifts. Moreover, the direct and indirect repercussions of U.S. trade restrictions may worsen the situation. Tariffs could be offset by further subsidies; the quota accords may crumble.

Most critically, the United States is not a bastion of pure free trade. Protectionism often has been employed where little evidence of foreign inefficiency existed. Continued protection of steel is more likely to encourage further protectionism abroad than to discourage it. Those of us who believe in the virtues of free competitive markets simultaneously deplore all the interventions that have occurred in steel and warn that nothing short of major economic reforms abroad will produce improvements.

Dr. Lima's study is an important contribution to our understanding of world steel markets, but there is much yet to be done. The Center for Energy and Mineral Policy Research has further projects under way and hopes to secure support for the extensive additional work that is needed. Nancy Warner did make a superhuman effort to transform the manuscript into camera-ready copy.

Richard L. Gordon

---

[2] Thomas R. Howell, William A. Noellert, Jesse G. Kreier, and Alan Wm. Wolff, *Steel and the State: Government Intervention and Steel's Structural Crisis*. Boulder: Westview Press, 1988.

Acknowledgments

Several people have been instrumental in the successful completion of this book. Of them, I would especially like to thank Dr. Richard L. Gordon for his helpful comments and expert guidance. Other useful advice was provided by Timothy J. Considine, William A. Vogely and James A. Miles of the Pennsylvania State University. A great debt is owed to people in the industry for providing valuable insights and data.

Finally, I wish to thank my wife, Catari, for her constant support and unwavering optimism.

Financial assistance was provided by the Conselho Nacional de Pesquisa e Desenvolvimento-CNPq and by the Companhia Siderurgica Nacional-Siderbras-Brazil.

Jose Guilherme de Heraclito Lima

1
Introduction

Since World War II, substantial changes have occurred in the world steel industry. This book deals with the possible consequences of these developments for steelmaking practices around the world, particularly in the United States. The central issue is whether it would be desirable for the United States and other industrialized countries to import steel slabs from newly established producers such as Brazil and South Korea. Pig-iron and semi-finished steel production seems cheaper in emerging countries than in countries that are already established steel producers. Producers in both types of countries can employ similar technologies and secure iron ore and coking coal at similar prices. Lower unit labor costs in the newly emerging producers outweigh capital-cost disadvantages.

The main issue discussed here is the emergence of slab imports. The available data on the comparative costs of such imports and of producing slabs in industrialized countries are reviewed. To develop the argument more fully, extensive background information is provided on critical developments within the steel industry.

Several basic changes require review. First, new steelmaking technology was, and continues to be, developed and widely employed. Second, new sources arose of critical raw materials such as iron ore and coking coal. Third, the level, location, and composition of steel consumption shifted drastically. These forces and others, such as the changing relative economic attractiveness of producing steel from pig iron or from scrap, profoundly altered production economics. Profitable opportunities arose to establish capacity in different parts of countries that already were major steel producers, and in countries that previously had not been significant producers. At least in the United States, new internal competition appeared from mini-mills, smaller mills that use electric furnaces to produce steel from scrap.

Radical shifts in the level and composition of demands for materials are another major influence on steel markets. Steel consumers throughout the world are seeking to produce lighter and more energy-efficient manufactured products. Shifting to alternative materials and lighter, higher quality steels are two ways to attain these product alterations.

Traditional steel-consuming markets also appear to be growing less rapidly than the overall economy. Low levels of private investment associated with both slow growth of high-capital-cost industries and the severe recessions that occurred during the the 1973-83 period further reduced world steel demand. Another influence widely thought to be important to steel market conditions is the belief that a domestic steel industry is critical to the development of an economy. This belief is alleged to produce uneconomic investment in steelmaking capacity.

Overcapacity prevailed in the world steel industry since 1975. World trade played an important and increasing role. The combination of weak markets and increased domestic and foreign competition has caused problems for the established steelmaking companies of the United States and Western Europe. The production trends, operating rates, imports, and exports of different countries also changed.

Lower consumption growth led to reduced steel production in such traditional centers as the United States and Western Europe. However, steel production increased elsewhere. Some of these changes occurred to accommodate new markets, but the new producers also developed exports to the United States and Western Europe. Predictably, this increase in competition has led to demands for protection, and in many cases, these demands have been met.

Since 1968, the U.S. steel industry, for example, has been protected by trade barriers. The barriers have been rationalized by standard complaints about unfair competitive practices employed abroad. In addition, the protection allegedly would provide a respite during which the industry would restructure its capacity and become competitive with imports.

An additional complication in the U.S. market is that the changing level, location, and composition of consumption, new technology, and changing economics radically altered the nature of steel production. In particular, the "integrated" producers, which traditionally produced the vast majority of steel output, lost markets to mini-mills. As discussed more fully in Chapter 2, integrated producers make a full range of steel-mill products in plants that produce pig iron as a main input to steelmaking. Integrated mills use scrap as a supplementary input to steelmaking.

Mini-mills, in contrast, produce a narrower range of products in plants that rely entirely on scrap for steelmaking. These mini-mills are more successful at producing products that do not involve flat-rolling. Flat-rolling technologies are most economic at larger scales than the mini-mills maintain and require the higher quality steel producible from pig iron.

U.S. integrated producers face the pressures of reduced consumption, import competition, and mini-mill competition. Protection has not sufficed to overcome these problems. As a result, integrated producers endured heavy losses, which in many cases caused bankruptcy. Even those with lesser difficulties found it desirable to dismantle capacity. Some plants closed; others had their capacities reduced.

This book, in effect, tests the applicability to steel of observations about the industrial structure made by George Stigler. He argued that heavy

Introduction 3

vertical integration is more likely when markets are small. As markets expand, open market trade in commodities also can expand. This, according to Stigler, is a major implication of Adam Smith's famous title of the third chapter of the *Wealth of Nations* : "That the Division of Labour is limited by the Extent of the Market."[1]

Stigler's principle has been implicitly explored by many writers apparently unfamiliar with Stigler's work. In particular, the whole literature on the commoditization of energy and minerals constitutes an implicit review of the Stigler hypothesis. Commoditization is another term for substituting open market trading for integration.

This book, in turn, is an extension of the commoditization literature to examine whether such a substitution is efficient for the steel industry. Verification of the basic hypothesis tested here is equivalent to proof that Stigler's prediction applies to the steel industry. While Stigler probably was stressing market expansion due to growing consumption, expansion also can involve increased linkages of previously geographically isolated markets. This book shows that such an increase in linkages has occurred in steel.

More fundamentally, if the hypothesis is correct, changes in U.S. trade policy could benefit the steel industry as well as the rest of the economy. Protectionism in steel almost certainly is bad for the economy. The literature on international trade shows that protection is beneficial to a country only under special circumstances. Few observers of the steel market believe that its characteristics allow the U.S. economy to gain from protection.

In particular, subsidies abroad are generally a source of benefits to other economies. Although the subsidization will harm U.S. producers, such damages to producers are less than the benefits to consumers. In the case of imported semi-finished steel, such imports may benefit, not only final consumers of steel, but also steel companies fabricating finished steel from imported slab.

Since protection hinders the development of trade in slab, it prevents survival and expansion of finishing mills that can operate profitably only if imported slab is substituted for domestically produced slab. The loss of finished-steel production might outweigh the benefits to the steel industry of protecting slab production.

AN OVERVIEW OF THE BOOK

Given the scattered nature of the data on steel-market conditions, this book provides extensive background material to place the role of slab-steel imports in a proper context. Much of this material is of interest only to those wishing extensive review. To facilitate examination of the most critical arguments, most of the details are concentrated in Chapters 3, and 5, and in

---

[1] Spelling and capitalization follow the 1979 version in the Glasgow Edition of the works of Adam Smith, published in the United Kingdom by the Oxford University Press and reprinted in the United States by Liberty Classics.

the two first appendices. The changes in market conditions reported here reflect economic and technological forces too complex to explain adequately within the confines of this book. The focus is on the implications for slab trade.

In this chapter, the objectives and structure of the book are introduced. In Chapter 2, technological developments in steel processing and the impact of these technologies on integrated producers are reviewed. Chapter 3 provides a broad discussion of key developments in the world and U.S. steel industries. First, the impacts of world excess supply and of the growth of international trade on industrialized, developing, and centrally planned economy countries are viewed. Second, the chapter notes consumption, production, and international trade trends in the U.S. steel market. Third, developments in the production and trade of iron ore, scrap, pig iron, and semi-finished steel and their implications are surveyed. Given the complexity of the material, the chapter summary highlights the critical changes.

Chapter 4 extends the discussion with information on shutdowns of U.S. steel-producing capacity. Chapter 5 then focuses in more detail on the markets for flat-rolled products and the role of slab imports in flat-rolled product production. Particular attention is given to the decline of U.S. integrated producers' capacity to produce crude steel. Steelmaking capacity is less than finishing capacity. Thus, full utilization of finishing capacity is possible only if alternative supplies of steel can be obtained. The economics of alternative patterns of steel supply are assessed in Chapter 6.

Chapter 7 reviews U.S. protectionism on the basis of the information developed in earlier chapters. Finally, Chapter 8 summarizes the main findings of the preceding chapters and presents conclusions on the basic hypothesis about the location of steel production and the implications for U.S. trade policy.

Appendix A shows the U.S. industry' financial plight. Appendix B lists the plant closings noted in Chapter 4. Appendix C has capacity balance equations for Chapter 5. Appendix D has the cost estimation procedure for Chapter 6.

2

Existing and Emerging Steelmaking Technologies and Their Market Implications

This chapter outlines the basic nature of existing steelmaking technologies and surveys prospective technological changes. Key facts about steel production processes are reviewed to facilitate understanding of the subsequent discussion. Information explaining the problems facing the industry and particularly those affecting integrated producers is stressed. First, an overview is provided. More detailed discussions focus on coking, methods to produce iron without using coke, steelmaking, and steel finishing.

The past developments in the utilization of technology critical for this book are those that permitted less integrated producers to compete more effectively. The prospective developments most relevant to the book are those that further increase the ability of non-integrated producers to compete and those that reduce the role of traditional iron-making processes.

AN OVERVIEW OF STEEL PRODUCTION

The basic distinction is between integrated and non-integrated producers. Three stages of steel production--iron-ore reduction, steelmaking, and steel finishing--prevail. Integrated mills undertake all three stages; non-integrated production usually bypasses iron-ore reduction and employs scrap as the main input into steelmaking.

For a long time steel production was dominated by integrated producers charging iron ore and coke in blast furnaces to reduce the ore to pig iron. Development of electric-furnace technology provided the alternative of making steel entirely from scrap. Non-integrated production consists mainly of finishing steel produced in electric furnaces making steel entirely from scrap. Non-integrated production consists mainly of finishing steel produced in electric furnaces.

Such non-integrated production has grown in importance in the United States since the 1960s and accounts for an important share of steel production. Producers termed "mini-mills" became important competitors. Such mini-mills have lower costs than do integrated producers, at least in producing some products. Thus, scrap use has substituted for manufacturing pig iron and for iron-ore consumption.

As population and industry centers shifted, steel consumption grew most rapidly in those regions of the United States where little efficient integrated capacity existed. When they arose around 1960, mini-mills pursued a strategy of locating near markets in which scrap supplies and local labor were available, but which were distant from integrated mills (Mueller, 1984). The transportation cost advantage over U.S. integrated producers enjoyed by mini-mills and some foreign suppliers contributed to their ability to compete in these emerging markets.

Around 1970, mini-mills began to compete more directly with integrated mills and adopted the strategy of continuously producing a limited number of high-volume steel products. Emphasis was on maintaining small-scale and relatively simple operations.

Most mini-mills are of very recent vintage, employing the latest technology in furnaces, continuous casters, and rolling mills. Indeed, mini-mills often refurbish or completely replace a production unit after only a few years of use. The mini-mills have no investment in raw materials or transport facilities and do not produce an intermediate product such as pig iron. They rely on others, such as scrap dealers and utility companies, for raw materials. Therefore, mini-mills can concentrate their investments on new fabrication technologies.

The 50 mini-mill plants are dispersed throughout the United States but are especially concentrated in southern states, in which the markets for mini-mill products grew because of new construction in the 1960s. In contrast, integrated mills are located largely in the Midwest, near raw material supplies, particularly of coking coal, and close to major industrial consumers (U.S. International Trade Commission, 1988b).

Elimination of blast furnaces also allows efficient operation at lower outputs than are optimal for integrated firms. At least two electric-arc furnaces (EF) are required to maintain operations, even when one of the furnaces is shut down for relining. Outputs of 250,000 to 600,000 tons per year can be efficiently produced from two electric-arc furnaces. Almost two-thirds of U.S. mini-mill capacity are in plants with a capacity of less than 600,000 tons per year (Barnett and Crandall, 1986).[1]

Optimal integrated plant capacity depends on the size of blast furnaces. The minimum efficient annual capacity of these furnaces is around 1.5

---

[1] The weights and measures used in this book are in metric system units except when noted.

million tons. Raw-material-processing and blast-furnace technology developments allow profitable operation of blast furnaces with more than 3 million tons of capacity. The need for shutdowns for blast-furnace relining requires that a plant have at least two blast furnaces. Thus, annual capacity must be in the 3 million to 6 million ton range. Actual capacity tends to be in the 4 million to 9 million ton range.

Coke is used in the blast furnace to support and reduce the iron ore. The output of the blast furnace is pig iron which serves as an input to steelmaking. Pig iron has a high carbon and silicon content, which provide energy when reduced by oxygen in the steelmaking stage.

For reliance on the blast furnace to continue, many economic problems must be solved (McManus, 1984a). However, technologies exist for reducing iron ore without the use of coke. Such technologies are termed "direct-reduction methods" because the intermediate step of coking is eliminated. The reducer in direct reduction can be any fuel, and the choice depends on the technology used and the comparative cost of different fuels. The product of direct reduction is sponge iron or direct reduced iron (DRI) that has a higher concentration of iron and less silicon and carbon than pig iron. DRI is normally transformed into steel in electric-arc furnaces. An intermediate degree of vertical integration can be achieved by building a direct-reduction facility to serve a mini-mill.

Whether DRI production is undertaken depends upon its cost relative to the costs of using scrap and of continuing reliance on blast furnaces. Such direct-reduction processes have accounted only for a small portion of world crude-steel production, but their adoption has increased (Innace, 1985a). In the United States, the price of scrap is lower than DRI production cost, and a clear cost advantage over the blast furnace is not apparent. Therefore, direct reduction production is small, and the U.S. steel industry is composed predominantly of integrated producers and mini-mills.

In the 1980s, molten pig iron is made into steel mainly in a basic oxygen furnace (BOF). Before World War II, the open-hearth furnace (OH) was the dominant technology for producing steel from pig iron. In the 1950s, the BOF began to replace the open hearth furnace because the BOF required much shorter heating times and less labor and capital per ton of output than the open hearth.

In the finishing stage, the liquid steel from steelmaking is cast by one of two processes. The metal can be "ingot cast," i.e., poured into iron molds and allowed to solidify into ingots. To produce semi-finished shapes, the ingots are heated to a temperature of 1000 to 1200 degrees Celsius in "soaking pits" and then rolled. In the other process, "continuous casting," the liquid steel is cast into an oscillating water-cooled mold, and the semi-finished steel shapes (slabs, blooms, or billets) are produced as a continuous strand.

The semi-finished shapes are divided among slabs, blooms, and billets that differ in cross section. Slabs comprise the wider (610 to 1,520 millimeters) and thinner (50 to 230 millimeters) shapes; a bloom has a height of 150 to 300 millimeters and an equal width; a billet has a height of 50 to 125 millimeters and an equal width (United States Steel, 1985). These semi-finished steel shapes are inputs for finishing processes.

8   Existing Steelmaking Technologies and Their Market Implications

After the removal of surface defects (conditioning), the slabs, blooms, or billets are heated to a temperature of about 1200 degrees Celsius and hot-rolled into various forms. Various additional finishing operations such as pickling (surface cleaning after hot-rolling), cold rolling, annealing (heat-treatment to produce specific properties to steel), and coating then produce different final steel products (United States Steel, 1985).

Blooms and billets are used to produce heavy and light non-flat-rolled products, respectively. Wire rods, bars, rails, and structurals (i.e., those forms critical for construction uses) are the main non-flat-rolled products. Slabs are used to produce flat-rolled products. Slabs are processed into plate through a reversing mill or into hot-rolled coils through a hot-strip mill. Plate is either sold or further processed into pipes. The hot-rolled coil can be cut into sheets and strips or may be further processed into cold-rolled products that could be sold as is or coated. Principal flat-rolled products are plates, hot-rolled sheets and strips, cold-rolled sheets and strips, and coated and electrical sheets (including galvanized sheets and tin-mill products). Hot-rolled coils can be semi-finished steel when re-rolled by steel producers but normally are sold to steel consumers as a finished product. Therefore, published statistics include hot-rolled coils among finished rather than semi-finished steel products.

Integrated firms can make both flat-rolled and non-flat-rolled products while mini-mills only produce various non-flat-rolled products. As discussed in Chapter 3, mini-mills captured large shares of the latter markets, and integrated producers concentrated more on flat-rolled products. The technology to produce flat-rolled products was economic only at the large scales attained by integrated producers. Since the technology necessary for efficient operation on a smaller scale is being developed, mini-mills may be able to compete in the flat-rolled products market (Isenberg-O'Laughlin, 1985a).

Since some countries lack iron ore, metallurgical coal, or both, they engage in international trade in iron ore and metallurgical coal. Also, international trade of iron ore and metallurgical coal is undertaken in countries in which total reliance on domestic production of these materials is uneconomical. Alternatives exist, such as the use of imported pig iron, merchant DRI, and scrap to produce crude steel. Another option is to trade in semi-finished steel shapes. The option of moving more processed products from one country to another involves a trade-off between the benefits of lower shipping costs and possibly higher processing and marketing costs. This is discussed in more detail in Chapter 6, in which the production costs in different countries are compared.

Conviction is widespread that blast furnaces and coke ovens are too big, too costly, and too inflexible (McManus, 1984a and 1985). Steel researchers are exploring processes that allow profitable operation at lesser scales than those of an integrated mill, and reduce investment costs per ton.

A final set of long-standing problems are associated with scrap. Existing conventional steelmaking technology imposes limitations on scrap usage. A BOF charge with a higher proportion of pig iron also can be more cheaply heated to the higher temperatures required in large slab casters. The BOF uses a maximum of 35 percent scrap charge. Using large quantities of scrap charge in the OH process lengthens the tap-to-tap time and is economically, and energy-inefficient. Therefore, the ways to increase scrap usage are to add more EF capacity and to develop new steelmaking technologies that allow more scrap-charging.

Newer difficulties are emerging with changes in the quality and quantity of scrap that may lessen its cost advantage over molten pig iron or DRI. Scrap is generated from steel-processing losses in steel production (home scrap) and steel transformation (prompt scrap) and by the disposal of consumer and durable goods (obsolete scrap). Home-scrap supplies tend to decrease as more continuous casting occurs. Steel consumers are adopting processes that reduce the availability of prompt scrap. Obsolete scrap tends to have more contamination than home and prompt scrap. The need to remove these impurities raises costs and makes obsolete scrap a less attractive input than home or prompt scrap. Such deteriorations in scrap-supply conditions could reduce the portion of steel output that could be efficiently produced from scrap.

COKING COAL AND COKE OVENS

Most integrated plants produce coke. Coals are blended and processed in a coke oven to provide suitable quality for the efficient operation of the blast furnaces in the production of pig iron. The blend can employ various mixes of coals.

Coking requires coals that possess the appropriate chemical and physical properties to agglomerate into a strong, low sulfur coke appropriate for steelmaking. Coals used in coking are termed metallurgical or coking coals. Concerns over metallurgical-coal supplies are perennial and largely ill-founded.

Historically, the scarcest, most expensive coals in the blend were those low in volatility content. Coals with medium and high volatile-matter content are in more ample supply and indeed can be obtained in substantial amounts by more thoroughly cleaning many of the low-sulfur bituminous coals used for steam raising.

Prices of metallurgical coals are higher than for other types of coal because of these quality requirements. It proved possible to reduce production costs by developing new sources of coking coals, particularly in Australia and Canada, and by lessening the role of the more scarce low volatile coals.

A short-lived increase in the price of metallurgical coal in the early 1980s inspired efforts to lessen dependence on traditional coking coals (Calarco, 1988). Technologies were developed to permit use of what are termed semi-soft coking coals, previously unsuited for coking (Calarco, 1988). The use of such semi-soft coking coals in manufacturing coke grew rapidly since 1980, reaching a level of 13.5 million tons in 1987. Of the total semi-soft coking coals consumed in 1987, about 2 million to 2.5 million tons were injected directly into the blast furnace (Calarco, 1988).

Technological advances also facilitated substitution of oil, gas, or uncoked coal for coke in the blast furnace. Prior to rises in oil prices in the 1970s, oil injection in the blast furnace was used to reduce coke consumption.

From 1973 to 1985, the average coke rate for a U.S. blast furnace fell from 597 to 525 kilograms of coke per ton of pig iron. In contrast, the Japanese coke rate was 426 kilograms in 1979. This low Japanese coke rate was achieved through the use of supplemental fuels. As oil use was phased out, the coke rate rose; it was 482 kilograms in 1986 (Tex Report Ltd., 1988). Since coke is cheaper in the United States, supplemental fuel may be less desirable for U.S. producers.

The development of pulverized coal injection directly in the blast furnace allows the use of coals generally not suited for coke manufacture. This technology was adopted in 18 of the 37 operating blast furnaces in Japan, the country most active in the installation of this process. Installation or experimentation with this technology took place in the United Kingdom, Sweden, Belgium, West Germany, the Netherlands, Italy, South Korea, Taiwan, Brazil, and some centrally planned-economy (CPE) countries. Pulverized coal injection has averaged 60 to 80 kilograms per ton of pig iron, but Kobe Steel in Japan achieved 100 kilograms per ton of pig iron (Calarco, 1988).

The adoption of fuel injection has several economic advantages. The substitution for each ton of coke eliminates the need for 1.4 to 1.6 tons of coal to produce coke. Investments in the injection process cost considerably less than building coke ovens. Blast-furnace control and productivity are improved, the heat content of blast-furnace gas increases, and steel producers can take greater advantage of changing relative prices of fuels.

ALTERNATIVE IRON-ORE REDUCTION PROCESSES

Steel producer interest in substitutes for the blast furnace increased since the early 1970s. The high capital costs of building blast furnaces and coke ovens and the costs of operating and maintaining those units provide incentives for developing new reduction technologies.

While many such technologies exist, discussion here can be limited to one particularly promising method, the Corex process. It replaces coke with coal in the production of pig iron with its quality at least equal to that produced from blast furnaces. The chemical composition and the tapping temperature of the Corex hot metal is equivalent to blast-furnace hot metal with the extra convenience of better control of silicon content. The recovery gas from the process has a higher heat value because a high-oxygen injection rate eliminates the presence of nitrogen.

The recovery gas is 95 percent carbon monoxide and can be used as a supplementary fuel in the blast furnace. The first commercial Corex plant was built by ISCOR in South Africa with a capacity of 300,000 tons of pig iron per year, and its economic feasibility should be known by late 1989 (Korf Engineering GmBH, n.d.).

Corex furnaces also can efficiently operate at a fifth of the scale of a blast furnace. The estimated capital expenditure for a Corex plant with an annual output of 300,000 tons is around $150 per ton. The cost of production of molten iron is $115 per ton when purchasing coal and iron ore at $50 per ton and $39 per ton, respectively (Korf Engineering GmBH, n.d.).

In 1988, the cost of investment in conventional technology to produce pig iron was $314 per ton, and the operating cost in an efficient integrated producer was $137 per ton. As a result, investment and production costs are reduced by 52 percent and 16 percent, respectively, when the Corex process replaces conventional technology.

In the meantime, techniques for repairing rather than replacing refractories are used to reduce the cost of maintaining blast furnaces. Full relining and rebuilding had to occur every four to six years. Refractory repairs may allow a blast furnace to operate for eight years before being fully rebuilt (Soares, 1987). This defers a cost ranging from $15 million to $100 million depending on the size of the furnace and the extent of rebuilding (U.S. International Trade Commission, 1988b). However, integrated steel producers still must eventually decide whether to rebuild their blast furnaces, and the deterioration occurring when a rebuild is delayed raises the cost of the subsequent rebuild.

STEELMAKING TECHNOLOGY

Research on steelmaking technologies has concentrated on increasing the ability to use scrap, improving control of product chemistry, and reducing energy consumption. Approaches include modification of the BOF to increase the proportion of scrap it can economically handle, reducing EF energy use, and developing steelmaking vessels that will be a hybrid of the BOF and EF.

One approach to allowing greater scrap use in the BOF is to increase energy input. The oxygen is the main supplement to the heat in the hot metal. Originally, the BOF process injected oxygen into the top of the furnace; some later processes employed blowing oxygen into the bottom of the furnace. Work is under way to develop techniques in which oxygen is injected both at the top and the bottom of the furnace so that more supplemental heat is available and more scrap can be used.

Klöckner's KMS process is the best proven technique available to steel producers seeking a flexible oxygen steelmaking vessel. The KMS, a combined blowing vessel, has been operating on an industrial scale since 1978 and allows scrap rates of 50 percent or higher. The KMS or related developments were installed throughout the world and produced more than 30 million tons per year of high-grade steel in 1985 (Innace, 1986).

However, the experience inspired questions about the cost of converting BOF shops to any combined blowing furnace. Steel experts argue that the cost of converting BOF to KMS can equal that of a complete blast-furnace reline ($80 million to $100 million). Nevertheless, Klöckner officials set pay-back, in some cases, at less than two years (Innace, 1986).

Ladle metallurgy facilities that enhance steel quality by removing impurities also facilitate scrap use by eliminating most, but not all, of the contaminants introduced as a result of scrap charging and melting. The main limitation is an inability to eliminate copper and tin. The technology also improves the quality levels because of its ability to adjust final chemistry and clean steel as close to the caster as possible, making high tonnages of BOF operation viable. Domestic steel producers, therefore, have invested in ladle metallurgy facilities, and about 13 systems have been installed in the United States (U.S. International Trade Commission, 1988b).

Technological developments for EF furnaces seek to reduce electric energy consumption per ton of molten steel and tap-to-tap time and to increase refractory life. Increasingly, the electric-arc furnace is being operated as a high-speed melting machine, reducing heat times from five hours to one and a half hours. Water cooling is one of the keys to this reduction in heat times, since high power inputs are possible without excessive refractory wear.

Another technological development entails using direct current (DC) instead of alternating current (AC) and results in a cost saving of $8 per ton. The cost saving is attained by using one graphite electrode instead of the three electrodes used in an AC-arc furnace and results in a 50 percent reduction of electrode consumption. Another benefit of a DC-electric furnace is lower furnace-floor noise. Such noise reduction decreases hazards to workers and facilitates closer worker attention to furnace operation. The investment cost is $800,000 per furnace, and the pay-back from reduced electrode consumption occurs in 18 months (McManus, 1985)

The Krupp and ASEA research program also includes the development of a DC-arc furnace for direct reduction of iron ores. Another concept being developed by Voest-Alpine is a plasma furnace that does not require electrodes. The high temperatures possible with plasma make it possible to attain direct reduction of iron ore (McManus, 1984a).

Development of new direct steelmaking technologies appears attractive to U.S. steel producers. The U.S. steel producers, through the American Iron and Steel Institute (AISI), are preparing a proposal asking the U.S. government to fund partially an estimated $20 million, three-year initiative to develop direct steelmaking technology.

If the experimental stage succeeds, an estimated $80 million or more would be needed to test the process at a full-scale plant (AISI, 1988). The Ling-Temco-Vaught (LTV) plant in Cleveland would be a leading candidate for that testing as the company already has invested in the project.

Justifications for the AISI initiative include the fact that direct steelmaking could lead to a $10 to $25 per ton variable operating-cost advantage over the coke-oven, blast-furnace, BOF route, reduced life-cycle capital costs, and elimination of coke batteries (AISI, 1988).

CONTINUOUS CASTING AND
FLAT-ROLLED FINISHING TECHNOLOGY

Although price is still critical when competing for markets, quality and services also are important because steel demand is very specific. Market pressures have increased the demand for improved steel quality. This requires producing steel with specific, homogeneous chemical and physical properties, greater gage, shape, and profile regularity, lower specific weight, and higher strength. Different kinds of coatings, metallurgical treatments, and lighter materials (high-strength, low-alloy steel and thinner gages) also are required.

Evidence of the role played by quality and service is found in a February 1986 report from the American Iron and Steel Institute. The report indicated that the automotive industry was working directly with integrated steel producers to improve the quality level of steel sheets for exposed automotive applications and delivery reliability of all sheet steel. The Ford Company decided that sourcing (contracting with suppliers) must be less subjective and more performance-oriented. The most significant changes in sourcing steel were:

--changing the weights given seller attributes in rating offers;
--raising the weight given quality and delivery performance from 50 percent to 80 percent (quality was weighted at 50 percent and delivery at 30 percent);
--reducing the weight given price/cost contributions from 50 percent to 20 percent;
--shifting from multiple to single sourcing of steel for an individual part; and
--sourcing semi-annually instead of annually to reward better performance sooner.

Automotive manufacturers reportedly are telling steel producers that the new robotic feed system requires sheets with precise quality. Allowable variation in thickness was halved from that previously permitted, and a reduction to a quarter of the old amount is expected.

The demand for higher-quality products is attributable to the consumer characteristics of buying the input on a weight basis and selling their output on an area basis. The demand for low-specific weight and high-strength steel resulted from substitution of other materials for steel. The need for homogeneity of properties is tied directly to increased process automation by steel consumers.

This means that it will be more difficult to produce large quantities of steel at standard gages and qualities. Producers need rolling mills and finishing facilities that will satisfy these new requirements. The shift to the use of smaller amounts of higher quality steel requires that mills operate efficiently at lower scales than those maintained by traditional integrated plants.

The change in the preferences of steel consumers has encouraged the adoption of quality-increasing technologies. Aside from a better steel refining process, such as ladle metallurgy, better performance was demanded from casting and finishing equipment. The improvements in casting and finishing involve providing more homogeneity in chemical characteristics and in the gage, shape, and profile of steel products.

Steel producers seek to reduce production costs and to meet demands for improved product quality. Adoption of continuous casting contributes to both goals. The cost-reducing benefits include a shorter sequence of operations, increased yield, improved product quality, energy savings, cheaper pollution control, and labor cost savings. Metallurgical quality improvements include less variability in chemical composition and solidification characteristics.

Both capital and operating costs are lower with continuous casting than with ingot processing. Capital cost savings arise from elimination of equipment required for ingot processing. Operating cost savings are primarily the result of lower manpower, energy requirements, and higher yields. Operating cost reduction can range from $30 to $35 per ton *(World Steel Dynamics*, 1987b). Pollution reduction is attained through the elimination of semi-finished steel scarfing and combustion from soaking pits.

In continuous casting, the surface quality of the cast is superior to that of a semi-finished steel produced from ingots. Consequently, continuously cast products can continue to be produced without any conditioning and can be hot charged in re-heating furnaces, thus reducing the energy consumption of re-heating furnaces by 30 percent to 40 percent.

The key to regularity of gage, shape, and profile lies in the hot-strip mill. In the hot-strip mill, profile irregularities can be corrected without introducing flatness problems and gage irregularity, which is critical for further processing.

The main technologies that produce the quality improvements required by steel consumers are sensing devices that provide quick response to dedicated process computers to correct problems, better rolling technologies, and devices to minimize energy losses.

The adoption of improved technology allows for the integration of rolling facilities on a continuous basis, which has introduced a major revolution into the steel industry. For example, the time between ordering and delivering a cold-rolled sheet, which took 40 days with conventional technology, falls to about 3 days. As a result, material-in-process inventories decline (Soares, 1987). U.S. hot-strip mills are undergoing major renovations to attain the cost reductions and quality improvements required by steel consumers.

Perfecting new casting technologies, such as for a thin-slab caster, the strip caster, and the thin-strip caster, is another area of extensive effort. Success could radically reduce both costs and the scale at which flat-rolling can be efficiently conducted.

The goal of thin-slab casting is to link molten-steel casting directly with the finishing stands of the hot-strip mill, eliminating re-heating furnaces and the rougher stands of the hot-strip mill. The objective of the strip caster is to remove completely the need for the hot-strip mill and re-heating furnaces. The thin strip caster would produce the cold-strip product directly from the casting of molten steel. A thinner continuous-casting product leads to a lower cost per ton. However, reduction of thickness can lead to changes in product quality that are unacceptable to many customers.

In the United States, four groups--USS and Bethlehem, Armco and Westinghouse, National and Battelle Columbus Labs, and Nucor and ASEA--are developing thin-slab caster technology.

Although the approach of each research effort differs, it is anticipated that the capital cost savings will reach $39 per ton and the operating cost savings will reach $50 per ton (Isenberg-O'Laughlin, 1985a). While direct steelmaking compresses the integration chain, the continuous casting of thin slabs will work against integration by making it more practical for nonintegrated mini-mills to make sheet products. Nucor, in its plant in Darlington, South Carolina, most likely will be the first steel producer to test the industrial application of thin-slab casting. Start-up is scheduled for 1989. Nucor considers a 500,000-ton plant possible. A conventional slab caster requires a minimum of 1 million tons capacity *(33 Metal Producing,* 1986).

Therefore, the development of thin-slab casting coupled with the other developments in steelmaking technology could conceivably allow mini-mills to produce flat-rolled products. Steel-production flexibility would increase; capital and operating costs would fall; sensitivity of profits to operating rates would decline.

SUMMARY AND CONCLUSIONS

By definition, technical progress involves improving products and reducing unit production costs. Substantial, successful efforts were expended on steel-industry technology. The traditional large integrated firm transforms coal into coke, combines the coke with iron ore to produce pig iron, combines pig iron with scrap to produce crude steel, and fabricates the crude steel into the shapes actually sold. The technology at every stage of this process has improved. The most striking changes were the success of three radically new technologies--the BOF and EF for steelmaking and continuous casting for finishing. Further developments, such as developing a more economic way to produce iron without using coke, are likely.

A key characteristic of new technology is that it produces unexpected results. In the case of steel, the combination of options made possible by new technologies, changes in raw-material-supply conditions, and shifts in the level, location, and composition of steel consumption, have altered world production practices. This chapter suggested that for certain products mini-mills were strong competitors in the United States. The rest of this book examines shifts in world production and consumption patterns. Particular attention is given to the possibility that one or more of the steps between iron production and steel finishing would be conducted in iron-ore-rich countries rather than in consuming countries.

3

World and United States Markets for Steel and Steelmaking Raw Materials

Changes in markets for steel and the inputs for steelmaking produced semi-finished steel-trade opportunities. This chapter reviews steel consumption, production, international trade and capacity utilization trends.

Trends in industrialized, developing, and centrally planned-economy (CPE) countries are compared. The noncommunist countries are subdivided further on the basis of market and raw-material-supply conditions. More detailed data are presented on steel in the United States. Finally, developments in raw materials for steelmaking are viewed.

The countries classification is mainly that of the International Iron and Steel Institute (IISI). Two exceptions were necessitated by the form in which the data employed here was compiled. Only the first nine of the twelve members of the European Economic Community (EEC) are covered. South Africa is included among the developing countries.[1] Differences in data availability force covering somewhat different time periods in different parts of the reviews.

The discussion is quite detailed, but the summary and conclusions section of this chapter presents the broad trends.

CONSUMPTION AND PRODUCTION TRENDS AND WORLD STEEL EXCESS CAPACITY

Since 1973, the rate and pattern of growth for materials, including steel, changed fundamentally. Steel demand leveled off, due to such influences as the changing composition of gross national products, the substitution of other materials for steel and its more efficient use (Larson, 1986). This stagnation in steel demand, combined with the growth of crude-steel

---

[1] The three latest members--Greece, Spain, and Portugal--are excluded. The countries included are Belgium, Denmark, France, West Germany, Ireland, Italy, Luxembourg, the Netherlands, and the United Kingdom.

production in previously non-steel producing countries, created excess steel-production capacity.

In 1979, world crude-steel production peaked at 747 million tons and then declined (Table 3.1). After falling to 646 million tons in 1982, recovery began, and from 1984 to 1987, output was over 710 million tons.

Overall, steel output of industrialized countries declined. Rising production prevailed in developing noncommunist and CPE economies (Table 3.1). From 1973 to 1986, world crude-steel production increased at an annual rate of 1 million tons. CPE countries' crude-steel production rose at an average annual rate of 6 million tons; that of developing noncommunist countries' crude-steel production, at an average annual rate of 4 million tons.

In noncommunist industrialized countries, steel production dropped at an average annual rate of 9 million tons, and the industrialized noncommunist countries' share of world crude-steel production fell from 65 percent to 48 percent. The developing countries' share increased from 5 percent to 12 percent; the CPE share went from 30 percent to 40 percent.

Crude-steel production in industrialized countries peaked in 1973-74 and then tended to diminish. Significant reduction of crude-steel production prevailed in the United States, Japan, and the EEC. Their combined share of world crude-steel production went from 58 percent to 40 percent. From 1973 to 1986, U.S. crude-steel production fell at an average annual rate of 5 million tons. Japanese and total EEC crude-steel production dropped at an average of 2 million and 3 million tons per year, respectively. Steel production in such industrialized countries as Italy, Spain, Portugal, and Turkey rose by an average of about 200,000 tons per year.

The leading examples of developing countries with expanded crude-steel production are Brazil and South Korea; crude-steel production in both grew at an average annual rate of one million tons. Crude-steel production in South Africa, India, Taiwan, Mexico, and Venezuela increased at an average annual rate of 200,000 tons. These seven countries' share of world crude-steel production increased from 4 percent to 10 percent. Production in the U.S.S.R. and China increased at an annual average rate of 2 million tons. North Korea, Rumania, Poland, East Germany, and Czechoslovakia increased their production by 200,000 tons per year.

World apparent steel consumption also expanded in the CPE and in developing noncommunist countries, and contracted in industrialized noncommunist countries (Table 3.1), where the average annual decline in apparent steel consumption equalled the 9 million ton-drop in crude-steel production. The CPE countries' apparent steel consumption increase, at an average annual rate of 7 million tons, exceeded the crude-steel production increase at an average annual rate of 6 million tons.

Conversely, developing countries' apparent steel consumption grew at an average annual rate of 3 million tons--one million tons below the 4 million tons production rise. This implies, as is shown below, that the CPEs raised imports and that developing noncommunist countries increased exports.

The share of apparent steel consumption of the industrialized countries went from 62 percent to 44 percent, the developing countries' share expand-

Table 3.1. Apparent Steel Consumption and Crude Steel Production by Region--1973, 1979, 1980, 1983, and 1986

	1973	1979	1980	1983	1986
Apparent Consumption of Steel (million metric tons)					
U.S.	146.45	136.93	111.44	90.87	91.56
E.E.C.	132.11	119.09	109.69	90.53	93.68
Japan	94.74	82.56	82.87	69.05	72.88
Industrialized Countries	428.68	394.22	359.35	298.87	312.47
Developing Countries	53.43	90.47	95.51	89.19	102.00
CPE Countries	209.20	260.05	258.06	265.95	300.72
Total World	691.31	744.74	712.91	654.01	715.19
Crude Steel Production (million metric tons)					
U.S.	136.80	123.69	101.46	76.76	74.03
E.E.C.	150.09	140.28	127.84	108.67	112.15
Japan	119.32	111.75	111.40	97.18	98.28
Industrialized Countries	456.93	433.65	397.91	336.62	343.14
Developing Countries	33.60	63.57	66.10	70.73	88.96
CPE Countries	206.63	249.66	252.38	256.50	283.30
Total World	697.16	746.88	716.39	663.85	715.41

Source: International Iron and Steel Institute, *Steel Statistics Yearbook*, Various Years.

ed from 8 percent to 14 percent, and the CPE countries enlarged their share from 30 percent to 42 percent.

In industrialized countries, Japan, the EEC, and the U.S. apparent steel consumption respectively fell by average annual rates of 1, 3, and 4 million tons. The EEC lowered apparent steel consumption and crude-steel production by equal amounts. U.S. and Japanese crude-steel production dropped more than apparent steel consumption. The share of world apparent steel consumption for the United States went from 21 percent to 13 percent; the EEC's, from 19 percent to 13 percent; Japan's, from 14 percent to 10 percent.

Apparent consumption in industrialized countries increased only in Portugal, Turkey, and Yugoslavia. Apparent steel consumption in developing countries--except in Argentina, Zimbabwe, and other African countries--rose (IISI, 1987b). The largest growth was in South Korea, Brazil, India, Taiwan, Saudi Arabia, Mexico, and Iran (IISI, 1987b).

From 1973 to 1986, apparent steel consumption increased in all the CPE countries. The largest growth took place in the U.S.S.R. and China, each of which raised apparent consumption at an average annual rate of 3 million tons. Rumania's growth, at an average annual rate of 290,000 tons, was the next largest among CPEs. The share in world apparent steel consumption for the U.S.S.R. and China moved from 23 percent to 43 percent.

World steel capacity was 950 million tons in 1986 compared to 786 million in 1973. Thus, gross steel capacity expanded on balance at an annual rate of 13 million tons compared to the million tons per year rise in production (Table 3.2). Although world crude-steel production started to dropped after 1979, world gross steel capacity declines at an average annual rate of 7 million tons did not begin until after 1982.

Again, changes differed among regions. From 1973 to 1986, the gross capacity of the CPE countries increased at an average annual rate of 7 million tons--one million tons a year more than output rose. The capacity of developing noncommunist countries expanded at an average annual rate of 5 million tons--also one million tons a year more than output increased. The capacity of industrialized noncommunist countries grew at an average annual rate of 10 million tons until 1980, then fell at an average annual rate of 15 million tons. Thus, no net change occurred over the full period analyzed.

From 1973 to 1986, the share of CPE countries in world steel capacity rose from 29 percent to 34 percent. The industrialized countries' share dropped from 64 percent to 53 percent; the share of developing countries increased from 7 percent to 13 percent. Capacity in both the United States and the EEC fell at an average annual rate of 2 million tons. Japan and other industrialized countries expanded their capacity by 1 million and 2 million tons per year, respectively. Construction of green-field (new plant) capacity in the noncommunist world prevailed in developing countries and will continue (Table 3.3).

Differences in changes of production and gross capacity necessarily affect gross operating rates (the ratio of production to gross capacity). The operating rate reflects the capacity utilization and indicates capacity available for increased production. Table 3.2 shows gross operating rates for industrialized, developing, and CPE countries. From 1973 to 1986, the gross operating rate in industrialized countries fell from 91 percent to 68 percent, and the CPE countries experienced a slight reduction from 90 percent to 88 percent. The gross operating rate in developing countries rose from 65 to 73 percent.

The effective capacity concept developed by Paine Webber's *World Steel Dynamics* is another way to assess capacity utilization. Gross capacity figures published by such major steel organizations as the AISI indicate the theoretical maximum production level that can be achieved by steel producers with ideal conditions prevailing at all existing plants. Effective capacity is the estimated level of steel production that is feasible over a sustained period of high utilization. Errors of estimating the capacity and temporary surges can and do lead to periods of greater than 100 percent utilization of effective capacity.

Effective capacity excludes plants in their start-up period or those being refurbished. Therefore, effective capacity better reflects how much capacity is available for production in the short run. Table 3.4 lists the effective capacity and effective operating rates (the ratio of production to effective capacity) of the industrialized and the developing countries.

In 1973, the United States, the EEC, and Japan together represented 82 and in 1986, 69 percent of the noncommunist world's effective crude-steel capacity. Gross and effective operating rates fell in these three areas (Tables 3.2 and 3.4). The largest impact and the lowest level was observed in the United States. Such declines date to the early 1970s for the U.S., the EEC,

Table 3.2. World Gross Capacity and Gross Operating Rate by Region-- 1973, 1979, 1980, 1983, and 1986

	1973	1979	1980	1983	1986
Gross Steel Capacity (million metric tons)					
U.S.	141.0	141.6	139.0	136.2	116.1
E.E.C.	175.0	203.5	204.7	186.8	152.1
Japan	135.0	157.0	158.8	162.0	153.7
Industrialized Countries	503.9	577.0	578.9	563.3	504.3
Developing Countries	52.0	82.8	86.1	105.2	122.4
CPE Countries	230.0	280.0	292.5	307.0	323.1
Total World	785.9	945.4	957.5	975.5	949.8
Gross Operating Rate (percent)					
U.S.	97.0	87.4	73.0	56.4	63.8
E.E.C.	85.8	68.9	62.5	58.2	73.7
Japan	88.4	71.2	70.1	60.0	63.9
Industrialized Countries	90.7	75.2	68.7	59.8	68.0
Developing Countries	64.6	76.8	76.8	67.2	72.7
CPE Countries	89.8	87.4	86.3	83.6	87.7
Total World	88.7	79.0	74.8	68.1	75.3

Source: *World Steel Dynamics*, "Steel Strategist #14," 1987.

and Japan, and to the late 1970s for the world as a whole. World excess steel capacity and low operating rates resulted.

INTERNATIONAL STEEL TRADE

Steel is traded in a growing, increasingly global market. Except for a fall in 1975, international steel trade expanded consistently. Actual international steel trade reached 156 million tons in 1986. Given conversion losses, the 1973 trade represented the equivalent of 170 million tons of crude-steel production--22 percent of the total. 1986 trade required 216 million tons of crude-steel, 30 percent of output (IISI, 1987).

Exports from the whole noncommunist world rose from 97 million tons in 1973 to 134 million tons of steel in 1986. The respective CPE levels were 17 million and 22 million tons (Table 3.5).

The increase in CPE steel exports was somewhat slower than that in world trade, so the CPE share of world steel exports fell from 15 percent to 14 percent. In the noncommunist countries, the industrialized countries exported 93 million tons in 1973 and 113 million tons in 1986; developing countries exported 4 million tons in 1973 and 22 million tons in 1986. The industrialized countries' share of world exports declined from 82 percent to 72 percent; the share of the developing countries increased from 3 percent to 14 percent.

Noncommunist countries imported 89 million tons in 1973 and 117 million tons in 1986. The respective CPE figures were 19 million and 39 million tons (Table 3.5). The noncommunist countries' import expansion

Table 3.3. New Greenfield Steel Plants Planned for 1985-95

Plant	Country	Process Route	Capacity (million m. tons)	Completion (year)
Acominas	Brazil	BF/BOF/BM	2.00	1986
Bellara	Algeria	DR/EF/CC	2.00	1990-95
Ajaokuta	Nigeria	BF/BOF/BM	1.30	1988
Misuarata	Libya	DR/EF/CC	1.26	1986-87
Nisic,Ahwaz	Iran	DR/EF/CC	1.50	1987
Nisic,Mobarakeh	Iran	DR/EF/CC	3.00	1989
Alexandria N.Steel	Egypt	DR/EF/CC	0.84	1986-87
Vizakhapatnam	India	BF/BOF/CC	1.20	1989
			2.20	1991-92
Daitari	India	BF/BOF/CC	1.30	1990-95
Posco,Kwanyang	S.Korea	BF/BOF/CC	2.80	1988-89
			2.80	1990
			2.80	1995
Perwaja	Malaysia	DR/EF/CC	0.58	1985

Note: BF means blast furnace.
DR means direct reduction.
BOF means basic oxygen furnace.
EF means electric furnace.
CC means continuous casting.
BM means blooming mill.
Source: *World Steel Dynamics*, "Steel Strategist #14," 1987.

was at a rate below that of total trade, so these countries' share of world steel imports dropped from 82 percent to 75 percent. The CPE countries raised their share from 18 percent to 25 percent.

Noncommunist industrialized countries imported 65 million tons in 1973 and 82 million tons in 1986. The respective developing countries' levels were 24 million tons and 35 million tons. The industrialized countries' share of world imports dropped from 60 percent to 53 percent, while developing countries maintained their share at 22 percent.

The main component of the slow rise in net exports from the industrialized countries from 1973 to 1986 was a fall in exports and an expansion by imports of the United States. The second most important change was a similar export decrease and import diminution for Belgium. The United States became a net importer, but Belgium remained a net exporter. Other industrialized countries generally maintained or raised exports while lowering imports.

The trend for other industrialized countries was that the net exporters maintained or expanded their net exports while net importers had smaller net imports. As a group, the industrialized countries were net exporters, and since 1974 maintained average net exports of 36 million tons (Table 3.5).

Developing countries raised their exports and contracted their imports. South Korean exports grew at an average annual rate of 400,000 tons, and

Table 3.4. Noncommunist World Effective Capacity and Effective Operating Rate by Region--1973, 1979, 1980, 1983, and 1986

	1973	1979	1980	1983	1986
Effective Steel Capacity (million metric tons)					
U.S.	133.9	130.3	127.9	125.3	106.8
E.E.C.	152.3	166.9	167.8	153.2	124.7
Japan	117.5	136.6	138.1	140.9	133.7
Industrialized Countries	449.7	495.9	497.2	484.4	433.6
Developing Countries	41.6	66.2	68.9	84.2	97.9
Noncommunist World	491.3	562.1	566.1	568.6	531.5
Effective Operating Rate (percent)					
U.S.	102.2	94.9	79.3	61.3	69.3
E.E.C.	98.5	84.0	76.2	70.9	89.9
Japan	101.6	81.8	80.7	69.0	73.5
Industrialized Countries	101.6	87.4	80.0	69.5	79.1
Developing Countries	80.8	96.0	95.9	84.0	90.9
Noncommunist World	99.8	88.5	82.0	71.6	81.3

Source: *World Steel Dynamics,* Steel Strategist #14, 1987.

imports grew at an average annual rate of 60,000 tons. In 1973, Latin America was a net importer, but by 1983 it had become a net exporter. Brazil's exports rose at an average annual rate of 400,000 tons while its imports fell at an average annual rate of 100,000 tons. Net exports by Argentina, Mexico, and Venezuela rose to a lesser degree (IISI, 1987).

In 1986, the developing countries in Asia were still net importers, although their steel production had grown. However, South Korea became a net exporter in 1979. All Middle Eastern and African countries (except Qatar and Zimbabwe) are net importers. The overall trend for developing countries was toward rising net imports through the 1973-81 period. Since then their net imports fell. In 1986, their net imports were 7 million below the 1973 level (Table 3.5).

From 1973 to 1986, the exports of CPE countries rose at an average annual rate of 400,000 tons. During the same period, their imports increased at an average annual rate of 1,500,000 tons. In particular, the U.S.S.R. and China increased their imports at an average annual rate of 1,100,000 tons and 300,000 tons, respectively. By 1973, the growth of imports by the U.S.S.R. and China made the CPE countries net importers. The CPE countries subsequently lifted their net imports to the extent that they have higher total net imports than the developing countries. By 1986, the CPE countries had net imports comparable to the United States (Table 3.5).

Table 3.6 shows data on countries having net imports or net exports above 3 million tons in 1986. Noncommunist countries represent 95 percent of the largest net exporters; the only CPE country considered to be a main net exporter is Czechoslovakia. In 1986, these main exporting countries accounted for 39 percent of the steel traded internationally, Japan being the largest net exporting country.

Table 3.5. Exports, Imports, and Net Exports of Semi-Finished and Finished Steel by Region--1973, 1979, 1980, 1983, and 1986

	1973	1979	1980	1983	1986
Exports of Semi-Finished and Finished Steel (million metric tons)					
U.S.	3.75	2.65	3.76	1.10	0.86
E.E.C.	54.49	62.12	59.19	54.15	60.02
Japan	24.80	30.66	29.69	30.88	28.71
Industrialized Countries	93.41	112.38	109.93	106.29	112.65
Developing Countries	3.70	11.86	11.19	19.23	21.73
CPE Countries	16.84	19.76	19.86	20.02	21.49
Total World	113.95	144.01	140.98	145.53	155.87
Imports of Semi-Finished and Finished Steel (million metric tons)					
U.S.	13.40	15.89	13.75	15.20	18.38
E.E.C.	36.52	40.93	41.05	36.01	41.55
Japan	0.21	1.47	1.17	2.74	3.31
Industrialized Countries	65.15	72.95	71.36	68.53	81.98
Developing Countries	23.53	38.76	40.60	37.69	34.77
CPE Countries	19.41	30.16	25.54	29.47	38.91
Total World	108.10	141.87	137.50	135.69	155.65
Net Exports of Semi-Finished and Finished Steel (million metric tons)					
U.S.	-9.65	-13.24	-9.98	-14.11	-17.53
E.E.C.	17.98	21.19	18.15	18.13	18.47
Japan	24.59	29.19	28.53	28.13	25.40
Industrialized Countries	28.25	39.43	38.57	37.75	30.67
Developing Countries	-19.83	-26.90	-29.41	-18.47	-13.04
CPE Countries	-2.57	-10.40	-5.68	-9.45	-17.42
Total World[a]	5.85	2.14	3.48	9.84	0.22

Note: [a] Net represents discrepancies between import and export statistics.

Source: International Iron and Steel Institute, *Steel Statistics Yearbook*, Various Years.

Table 3.6 shows fewer leading importing countries than main exporting countries. Only three--China, the United States, and the U.S.S.R.--are identified as major net importers. In 1986, these three countries accounted for 25 percent of steel traded internationally and, excluding the U.S.S.R., their share amounts to 22 percent. Imports by China and the U.S.S.R. helped to reduce excess supply in the international market.

China and the U.S.S.R. chose to rely on imports rather than to increase output at the same rate as consumption. Excess capacity in the noncommunist world and the resulting opportunities to secure steel at lower prices increased the attractiveness of imports. Explanation of the forces causing China and the U.S.S.R. to take advantage of these import opportunities is beyond the scope of this book.

The prior discussion suggests that four low-cost producer countries--Brazil, South Korea, South Africa, and Rumania--had particularly large increases in steel production and exports (Table 3.7). Although data to analyze the capacity and cost of production in Rumania and South Africa are inadequate, the South Korean and Brazilian steel industries have the lowest

Table 3.6. Main Net Exporters and Net Importers of Semi-Finished and Finished Steel in 1986

Countries	Net Exports of Steel[a] (million metric tons)
Japan	25
Belgium/Luxembourg	8
West Germany	6
Brazil	6
South Korea	3
South Africa	3
Czechoslovakia	3
France	3
Spain	3
China	-18
U.S.	-17
U.S.S.R	-4

[a]Negative numbers denote import exceed exports.

Source: Compiled from International Iron and Steel Institute, *Steel Statistics Yearbook*, Various Years.

operating costs among noncommunist countries. High capacity utilization, cheap labor, and modern facilities are the major reasons for these low costs (see Chapter 6).

The example set by these four countries was followed in other developing countries such as Argentina, Mexico, Venezuela, and Taiwan. They too attained low production costs, modern capacity, and high operating rates. This expansion in the exports of the developing countries, as noted, created new competition for the industrialized countries--especially in the United States, Europe, and, most recently, Japan. This competition put downward pressure on price and encouraged steel companies in industrialized countries to restructure their steel capacity and lower costs (see Chapter 4 for data on the U.S. experience).

Another way to classify countries is on the basis of a combination of market size and endowment of iron ore and energy. Table 3.8 shows the crude-steel production, imports, exports, net exports, and apparent consumption of steel for the countries categorized on such a basis.

The noncommunist world is divided into big market countries (the United States, the EEC, and Japan), iron-rich industrialized countries (Canada, Sweden, and Australia), iron-rich developing countries (South Africa, Brazil, and India), oil-rich developing countries with iron-ore reserves (Algeria, Argentina, Venezuela, and Mexico), other oil-rich developing countries (Middle Eastern countries, Indonesia, Nigeria, and Egypt), and other countries. This division was used to assess the competitive position of countries with an abundance of iron-ore and oil resources and to determine whether parts of steel production are moving closer to raw materials and further from markets.

Table 3.7. Apparent Consumption, Crude Steel Production, and International Trade in Semi-Finished and Finished Steel by Brazil, South Korea, South Africa, and Rumania--1973, 1979, 1980, 1983, and 1986

	1973	1979	1980	1983	1986
Apparent Consumption of Steel (million metric tons)					
Brazil	8.54	13.01	14.47	9.62	15.59
South Korea	2.47	7.20	6.12	8.31	11.56
South Africa	5.64	6.76	7.50	5.62	5.83
Rumania	8.26	12.64	12.47	11.40	12.02
Total above	24.91	39.60	40.55	34.94	45.00
Crude Steel Production (million metric tons)					
Brazil	7.15	13.89	15.31	14.67	21.20
South Korea	1.16	7.61	8.56	11.92	14.60
South Africa	5.72	8.88	9.07	7.18	8.90
Rumania	8.16	12.91	13.18	12.59	14.30
Total above	22.19	43.29	46.11	46.36	59.00
Exports of Semi-Finished and Finished Steel (million metric tons)					
Brazil	0.43	1.48	1.51	5.13	6.16
South Korea	0.81	3.14	4.54	5.73	5.89
South Africa	0.63	2.21	1.69	1.74	3.16
Rumania	1.39	1.95	1.98	2.18	3.06
Total above	3.26	8.78	9.72	14.78	18.27
Imports of Semi-Finished and Finished Steel (million metric tons)					
Brazil	1.82	0.60	0.66	0.08	0.55
South Korea	2.12	2.73	2.09	2.12	2.85
South Africa	0.55	0.09	0.12	0.18	0.09
Rumania	1.49	1.68	1.28	0.98	0.78
Total above	5.98	5.10	4.16	3.36	4.27
Net Exports of Semi-Finished and Finished Steel (million metric tons)					
Brazil	-1.39	0.88	0.84	5.05	5.61
South Korea	-1.31	0.41	2.44	3.61	3.04
South Africa	0.08	2.12	1.57	1.56	3.07
Rumania	-0.10	0.27	0.71	1.20	2.28
Total above	-2.72	3.68	5.56	11.42	14.00

Source: International Iron and Steel Institute, *Steel Statistics Yearbook*,

A 1977-to-1986 restructuring in the big market countries involved the decline of crude-steel production from 342 million to 285 million tons. The drop in apparent consumption was similar, and net imports in 1986 were only one million tons above 1977. The behavior of each of the three parts of the big-market-country sector was discussed above. The main point to recall from that discussion is that the change in net exports is the net effect of a rise of almost 12 million in net EEC exports, a fall of about 8 million tons in Japanese net exports, and a 2 million-ton increase in U.S. net imports.

From 1977 to 1986, iron-rich countries expanded their crude-steel production from 54 million to 68 million tons. The largest growth of crude-steel production took place in iron-rich developing countries, which

Table 3.8. Apparent Steel Consumption, Crude Steel Production, and International Trade of Semi-Finished and Finished Steel by Regions[a] of the Noncommunist World--1977, 1979, 1980, 1983, and 1986

	1977	1979	1980	1983	1986
Crude Steel Production (million metric tons)					
Big Market [b]	342.26	375.72	340.69	282.61	284.46
Iron-Rich Industrial	24.94	28.91	27.73	22.72	25.47
Iron-Rich Developing	28.56	32.90	33.89	32.09	42.33
Total Iron-Rich	53.49	61.80	61.62	54.81	67.79
Oil-Rich Dev. with Iron Ore	13.14	16.73	17.92	13.58	13.86
Other Oil-Rich Dev.	3.03	3.35	3.48	4.26	5.82
Total Oil-Rich Dev.	16.17	20.07	21.40	17.84	19.67
Other Noncommunist Countries	31.41	39.63	40.30	52.10	60.19
Noncommunist World	443.33	497.22	464.01	407.35	432.11
Exports of Semi-Finished and Finished Steel (million metric tons)					
Big Market [c]	89.04	95.43	92.65	86.12	89.58
Iron-Rich Industrial	6.51	6.76	7.21	6.60	7.51
Iron-Rich Developing	3.60	3.75	3.25	6.90	9.34
Total Iron-Rich	10.11	10.51	10.46	13.49	16.85
Oil-Rich Dev. with Iron Ore	0.52	0.92	0.59	2.53	2.92
Other Oil-Rich Dev.	0.19	1.32	0.62	0.58	0.69
Total Oil-Rich Dev.	0.71	2.24	1.21	3.11	3.61
Other Noncommunist Countries	9.24	16.07	16.80	22.79	24.34
Noncommunist World	109.09	124.24	121.12	125.51	134.38
Imports of Semi-Finished and Finished Steel (million metric tons)					
Big Market [c]	64.32	58.28	55.96	53.96	63.24
Iron-Rich Industrial	4.26	4.67	3.95	3.71	4.60
Iron-Rich Developing	1.62	2.57	2.61	1.99	2.64
Total Iron-Rich	5.88	7.24	6.56	5.70	7.23
Oil-Rich Dev. with Iron Ore	4.24	5.41	6.33	3.04	2.17
Other Oil-Rich Dev.	11.71	15.65	16.42	16.42	11.60
Total Oil-Rich Dev.	15.95	21.05	22.75	19.46	13.77
Other Noncommunist Countries	24.85	25.13	26.70	27.11	32.50
Noncommunist World	111.00	111.71	111.96	106.22	116.74
Apparent Consumption of Steel (million metric tons)					
Big Market [b]	317.55	338.57	304.00	250.45	258.12
Iron-Rich Industrial	22.69	26.82	24.47	19.83	22.55
Iron-Rich Developing	26.58	31.71	33.25	27.18	35.63
Total Iron-Rich	49.27	58.53	57.71	47.01	58.18
Oil-Rich Dev. with Iron Ore	16.86	21.21	23.66	14.08	13.11
Other Oil-Rich Dev.	14.56	17.68	19.28	20.10	16.73
Total Oil-Rich Dev.	31.41	38.89	42.94	34.18	29.84
Other Noncommunist Countries	47.02	48.70	50.20	56.42	68.35
Noncommunist World	445.25	484.69	454.85	388.06	414.47

Table 3.8 (continued)

	1977	1979	1980	1983	1986
Net Exports of Semi-Finished and Finished Steel (million metric tons)					
Big Market [c]	24.71	37.14	36.69	32.16	26.34
Iron-Rich Industrial	2.25	2.09	3.26	2.89	2.92
Iron-Rich Developing	1.97	1.18	0.64	4.91	6.70
Total Iron-Rich	4.22	3.27	3.91	7.80	9.62
Oil-Rich Dev. with Iron Ore	-3.72	-4.48	-5.73	-0.51	0.75
Other Oil-Rich Dev.	-11.52	-14.33	-15.80	-15.84	-10.91
Total Oil-Rich Dev.	-15.24	-18.82	-21.54	-16.35	-10.16
Other Noncommunist Countries	-15.61	-9.06	-9.90	-4.32	-8.16
Noncommunist World	-1.92	12.54	9.16	19.29	17.63

Notes: [a] See text for definition of regions.
[b] See Table 3.1 for details on U.S., EEC, and Japan.
[c] See Table 3.5 for details on U.S., EEC, and Japan.
Source: International Iron and Steel Institute, *Steel Statistics Yearbook*, Various Years.

expanded their crude-steel production from 22 million to 34 million tons. Crude-steel production in iron-rich industrialized countries only grew from 32 million to 34 million tons.

Iron-rich developing countries increased their apparent consumption of steel and reduced their dependence on steel imports. From 1977 to 1986, iron-rich developing countries raised their net exports from 2 million to 7 million tons. The growth of crude-steel production in iron-rich industrialized countries matched a net steel export rise from 2 million to 3 million tons. From 1977 to 1986, iron-rich industrialized countries substituted their own production of iron ore for imported iron ore when their steel production went from 25 million to 26 million tons.

The oil-rich developing countries had and continue to have plans to expand their share of world steel output. Steelmaking is based on direct reduction, electric furnaces, and continuous casting. Natural gas associated with oil production in these countries is a waste product. Therefore, the countries have sought to introduce industries that could employ the gas. Building plants for direct reduction of iron ore using gas was considered one attractive alternative, and a large number of such facilities were planned and built. The potential for further developments depends upon how gas supply-demand balances alter with energy-market developments and the completion of gas-consuming facilities.

In the 1970s and early 1980s, large oil revenues ensured a ready source of government financing for these projects. Over the same period, some countries indicated their intention to build direct-reduction plants with the sole purpose of exporting merchant DRI. Since then, merchant DRI became a substitute for scrap, which would affect the scrap market and constitute an initial step toward the manufacture of semi-finished steel products. While some of these plants were completed, they did not play a major role until 1986.

The oil-rich developing countries raised their crude-steel production from 16 million to 20 million tons. However, the oil-rich developing countries with iron-ore reserves only expanded their crude-steel production from 13 million to 14 million tons. Other oil-rich developing countries elevated their crude-steel production from 3 million to 6 million tons.

From 1977 to 1984, the oil-rich developing countries with iron-ore reserves were net importers of steel. Those oil-rich developing countries with iron-ore reserves increased their iron-ore production to produce more steel domestically and to satisfy their domestic demand without resorting to imports. These changes made them net exporters of steel in the 1980s. Other oil-rich developing countries raised their steel production through the imports of iron ore to produce more steel domestically and decreased their dependence on steel imports. Other oil-rich developing countries reduced their net imports of steel from 12 million to 11 million tons.

From 1977 to 1986, other noncommunist countries boosted their crude-steel production from 31 million to 60 million tons. As a result, their net steel imports declined from 16 million to 8 million tons. This growth of crude-steel production satisfied the rapid growth of demand and reduced dependence on steel imports. Apparent consumption of steel by the other noncommunist countries expanded from 47 million to 68 million tons.

THE UNITED STATES STEEL MARKET

Increased international steel trade and world excess capacity most severely affected industrialized countries, with the greatest impact on the United States. The U.S. steel market has been in a long-term decline and undergoing a major structural change. Between 1950 and 1986, the U. S. share of world steel exports fell from 20 percent to 1 percent, while its world steel imports' share grew from 9 percent to 17 percent.

U.S. Consumption

U.S. steel consumption grew during the early postwar period. Apparent consumption reached an all-time high in 1973 (while this was the year of the first oil shock, the exact contribution of oil price rises to changing material use patterns is unknown).

Shipments to those markets of the steel industry that purchase more than one million tons of steel products comprise more than 80 percent of sales. Thus, a review confined to such sectors conveys most of the essential information about steel demand. The main gap in the data used is that two groups of recipients reported are not final users. The more important is the service-center sector.

Service centers purchase steel from producers, warehouse it, and resell it, with or without such minor fabrication as slitting, shearing, drilling, and bending. The other reselling sector consists of converting and processing firms that engage in further finishing steel products. The data do not show to whom these two sectors sell, but service centers are likely to supply smaller-scale consumers that cannot be economically served directly by steel mills. The main categories of final consumers include the automotive, appliance, other domestic and commercial equipment, construction, oil and gas drilling, machinery, and container industries.

Table 3.9 shows U.S. industry shipments to these markets. In 1973, the largest share of steel shipments was to the automotive market, followed by those to service-center, construction, and machinery markets. These four markets represented 43 percent of U.S. industry shipments. By 1987, service centers were the main market for shipments from the steel industry, and the four main markets represented 61 percent.

A larger share of steel products previously shipped directly to end-user markets is transhipped through the service centers. The increased share of the service center can be attributed to consumers' increased need for services not provided by steel producers. In part, the shift involves integrated producer abandonment of labor-intensive activities to non-union firms. The change also reflects increased specialization by steel firms. They no longer act as a supermarket to their customers but supply a limited product range to a distributor who handles the products of a number of producers. These service centers are taking on the characteristics of Japanese trading companies in financing and marketing the products of others.

In 1986, service centers received 25 percent of total steel shipments and 36 percent of total flat-rolled product shipments. The service-center share of flat-rolled shipments included 47 percent of total hot-rolled sheet shipments and 30 percent of both galvanized and cold-rolled shipments (*World Steel Dynamics*, 1987). For imported steel, the service centers handle an even higher share. In flat-rolled products the growth of service centers rose from about 23 percent of shipments in 1980 to almost 50 percent in 1986 (*World Steel Dynamics*, 1987).

The automotive industry is the largest single consumer of direct-mill shipments and the principal consumer of hot- and cold-rolled sheets and galvanized sheets. In the years 1973, 1977, and 1978, when record numbers of motor vehicles were produced, the automotive industry consumed large quantities of steel. In 1979, automotive production reached an all-time high of 13 million units--9.3 million passenger cars and 3.7 million trucks. In that year, direct shipments to the automotive industry totaled 16.9 million tons, compared to 21 million tons in 1973. Shrinkage in vehicle size, substitution of other materials, and shifts to thinner steels made 1979 shipments lower than in prior years of high production.

Since 1979, the U.S. automobile industry faced a further shift in consumer demand away from its products because of intense competition from abroad. The automotive industry also altered its product lines to respond to the shift in demand to smaller cars. Total vehicle production fell to 8.0 million units in 1980 and 1981 and to 7.0 million units in 1982 (*U.S. Industrial Outlook*, 1986).

Given the sharp drop in production, 1982 shipments of steel to the automotive industry declined to a low of 8.4 million tons, which was 21

Table 3.9. Net U.S. Steel Mill Products Shipments to Main Sectors[a] -- 1973, 1979, 1980, 1983, 1986, and 1987

	1973	1979	1980	1983	1986	1987
Automotive	21.06	16.89	11.03	11.17	10.78	10.29
Appliances, Utensils & Cutlery	2.49	1.94	1.56	1.47	1.50	1.48
Construction & Contractors' Products	16.20	12.45	11.02	9.05	9.62	9.99
Service Centers & Distributors	18.48	16.54	14.67	15.17	15.85	18.00
Oil & Gas Industry	2.48	3.38	4.87	1.17	0.93	1.35
Containers & Packaging	7.08	6.14	5.03	4.11	3.72	3.97
Machinery, Industrial & Electrical Equipment	4.22	7.97	8.00	4.18	3.75	3.80
Other Domestic & Commercial Equipment	1.80	1.88	1.54	1.24	1.06	1.04
Converting & Processing	4.28	4.59	3.73	3.99	5.11	6.53
Total Above Markets	82.67	71.81	59.81	51.73	52.37	56.87
Total Shipments	101.04	90.97	76.10	61.31	63.76	69.57
Share of Above Markets from Total Shipments	81.80	78.90	78.60	84.40	82.10	81.70

Note: [a] Major sectors are those receiving more than 1 million metric tons in 1987.

Source: Compiled from American Iron and Steel Institute, *Annual Statistical Report*, Various Years.

percent of the steel shipments compared to 30 percent in 1977. Domestic motor vehicle production rose to 9.2 million in 1983; however, steel shipments recovered to only 11.1 million tons. Comparing 1983 with 1964 shows that although approximately the same number of vehicles were produced, steel shipments dropped by 5.7 million tons. Steel imports, however, were also a factor in this reduction *(U.S. Industrial Outlook*, 1986).

In addition to the down-sizing of cars, imported automobiles made up as much as 27 percent of U.S. automotive market *(U.S. Industrial Outlook*, 1986). Passenger car imports during the 1972-76 period reached 1.5 million units and by 1985 attained a range from 2.0 million to 2.4 million units. Truck imports in the early 1970s were in the 200,000-vehicle range but by the early 1980s had reached more than 400,000 vehicles. Although motor-vehicle production rose in 1984 and 1985 to 10.9 million units and 11.2 million units, respectively, it is unclear when the automotive industry will again reach a production rate near 13.0 million motor vehicles. Because of imports and smaller cars, automobile industry purchases of U.S. steel can be expected to be as much as 7 million to 9 million tons less than they were during the record years of the 1970s.

The average car weight in the United States fell from 1.54 tons in 1975 to 1.36 tons in 1985. Aside from the down-sizing of cars, materials substitution in the automotive industry played a major role. The amount of plain steel in a typical U.S. automobile fell from more than 1.135 tons in the early 1960s to less than 0.681 tons in 1985--from more than two-thirds to approximately one-half of the total weight of an automobile.

In the early 1960s, aluminium and plastics combined accounted for only about 1 percent of the weight of a typical automobile; however, in 1985, they accounted for well over 10 percent. The rise in the importance of high strength steel to the automobile industry was rapid. None was used in the early 1970s; more than 91 kilograms or 7 percent of the total weight was employed in 1986 (Eggert, 1986). This ability to adapt to new quality requirements illustrates the adaptation the steel industry had to make.

The construction market, which consumes most structural shipments, reduced its consumption of steel from 16.2 million tons in 1973 to 10.0 million tons in 1987. Although physical consumption dropped, the sector maintained the same share of U.S. industry shipments. Non-residential building, the principal steel-using part of the construction sector, declined (*U.S. Industrial Outlook*, 1986).

High freight charges cause the construction industry to seek nearby suppliers--typically ones less than 200 miles from the construction site. Domestic integrated steel producers are at a disadvantage here because they are located too far from the booming markets, thus allowing the mini-mills and imported steel to capture markets. Contracts for many sizable building and bridge projects on the West Coast and in inland areas were won by Korean and Japanese structural steel fabricators (Mueller, 1984).

From 1973 to 1987, machinery industries reduced their consumption of U.S. steel from 8.8 million to 4.2 million tons, and their share of U.S. steel shipments dropped from 9 percent to 6 percent. Rising machinery imports were a major element in the decline of shipments to the machinery sector. In 1973, the United States had a $2 billion net machinery-export sales balance, which rose to $5 billion in 1981. Since then, the net-sales balance fell as imports grew and exports dropped, and the 1986 net balance was only $0.5 billion. Imported machinery sales began to expand in 1976, remained constant until 1983, when they were $1 billion; in 1986, machinery imports were $2 billion (*U.S. Industrial Outlook*, 1986).

Another area in which steel markets shrank is the container market, which consumes most of the tin-mill products. Purchases fell from a high of 7.1 million tons in 1973 to 3.7 million tons in 1986, a drop of 3.4 million tons. The loss of market share is due primarily to inroads of other materials such as aluminum, plastics, impregnated paper, and glass. Aluminum captured the beer can market which consists of 20 percent of the entire can market, while other materials such as plastics and impregnated paper also captured markets from tinplate (*U.S. Industrial Outlook*, 1986).

Steel shipments rose for U.S. oil and gas production from 1973 (2.5 million tons) to 1981 (5.7 million tons) and then plummeted to less than 1.2 million tons by 1983. The 1981 figure was the result of a boom year when oil drillers were operating 4,530 rigs and leasing many of them at a rate of $12,000 to $14,000 per day (*Business Week*, 1986). Drillers did not want rigs to be idle for lack of pipe. Consequently, inventories of pipes were built-up, so that in early 1983, several million tons were on hand in oil-patch areas.

Drilling was expected to increase, and steel-demand levels were expected to remain at 3.0 million tons a year. However, crude-oil prices dropped dramatically in 1986, leading to less exploration activity and production (*Business Week*, 1986). The result was lower steel demand.

From 1973 to 1987, U.S. steel shipments to the appliance and other domestic and commercial-equipment markets went from 2.5 million to 1.5 million tons and from 1.8 million to 1.0 million tons, respectively. These sectors comprise the second largest consumer of steel sheets in the U.S. steel market. However, their share in total shipments through this period remained steady. One major reason for the decline in direct U.S. steel shipments to the appliance market was reliance on service centers. Material substitution also was a major influence (Larson, 1986).

Another growing market is the converting and processing sector. This market segment increased its purchases from 4.3 million to 6.5 million tons, which suggests that more firms are entering the manufacturing of steel to provide extra processing. Wire rods and hot- and cold-rolled sheet are the bulk of the shipments to this market.

Traditionally, total tonnage levels were accurate indicators of industry health. The increased stress on quality reviewed in Chapter 2 means success can be attained by lower volumes with higher quality and value. These changed customer requirements lessen the ability to maintain long production runs and the desirability of large-scale production. This, in turn, improves the position of mini-mills.

U.S. Production

1986 domestic crude-steel production was 74 million tons, 54 percent of the record level attained in 1973 (Table 3.10). Apparent steel consumption, after peaking at 147 million tons in 1973, fell to about 91 million tons in 1986. From 1973 to 1986, the 5 million tons annual average decline in crude-steel production was steeper than the drop in apparent consumption of 4 million tons.

During the same period, gross capacity dropped at an average annual rate of 2 million tons, a smaller reduction than that of apparent consumption and domestic crude-steel production. Gross operating rates, therefore, declined from 97 percent to 64 percent.

Given the high fixed costs of steel production, low capacity utilization hinders the recovery of fixed costs. One rule of thumb is that steel producers wish to maintain a gross operating rate above 80 percent. However, the operating rate necessary for survival actually varies with market conditions. Table 3.11 compares actual operating rates from 1981 to 1986 to the operating rates which major mills in the United States would have to attain to break-even. The operating rate at which break-even occurs necessarily varies inversely with price and directly with unit input cost. The fluctuations shown in this break-even rate presumably mainly reflect gyrations in prices.

Another problem of the U.S. steel industry arises from a slower rate of adoption of continuous casting. Metallic yields (the ratio of net shipments to crude-steel production) have not increased to the levels attained by countries such as Japan with greater reliance on continuous casting. Table 3.10 shows that the U.S. industry metallic yield rose from 60 percent in 1970 to 86 percent in 1987. In Japan, the 1985 metallic yield was 93 percent because most of the steel produced was continuously cast.

In 1987, crude-steel production and apparent steel consumption rose as steel demand rose. In the same year, gross steel capacity dropped to 102 million tons per year. As a result, the U.S. gross operating rate increased to 80 percent.

Table 3.10 shows that consumption declined less than production, and net imports grew. Table 3.12 shows manufacturers' shipments, exports, imports and apparent consumption by steel grade in the U.S. market from 1973 to 1987. Manufacturers' shipments of carbon and alloy steel dropped from 89 million to 67 million tons and 9 million to 6 million tons, respectively; carbon- and alloy-steel apparent consumption fell less than that of manufacturers' shipments. However, stainless-steel shipments tended to increase, but less than consumption. Net imports of all three steel types rose.

The predominance (over 90 percent) of carbon steel over alloy and stainless production, imports, and consumption persisted. However, the percent drop in carbon-steel exports exceeded that of alloy and stainless. From 1973 to 1987, exports of carbon-steel fell from 3 million to 1 million tons, while alloy- and stainless-steel exports were nearly the same as in 1973. As a result, the proportion of carbon steel in total exports dropped from 91 percent to 61 percent, and the share of alloy and stainless steel in steel exports rose from 7 percent to 24 percent and 2 percent to 7 percent, respectively. The share of alloy and stainless in imports grew by smaller amounts--from 2.5 to 5 and from 1.0 percent to 1.6 percent. The share of imports in apparent consumption increased for all three types but remained highest for carbon steel (Table 3.12).

Production efficiencies enabled mini-mills to dominate the production of reinforced bars, wire and related products, and light structurals. Mini-mills account for the majority of shipments of these products. From 1969 to 1986, mini-mills raised their share of bars and wire rod production (light non-flat-rolled products) from 12 percent to 67 percent and from 13 percent to 62 percent, respectively. The 1986 share in structural shapes was 62 percent.

As mini-mills captured a large share of the non-flat-rolled products markets, the share of flat-rolled products in the output of integrated producers rose. Table 3.13 shows the share of flat-rolled and non-flat-rolled products in the total shipments of the major integrated producers in 1981 and 1985.

A four-month United Steel Workers (USW) strike in 1959 was considered a major influence in stimulating steel imports and mini-mill supplies. During 1958-59, imports doubled from 2 million to 4 million tons and became a source of hedging against further strikes. After this strike, mini-mills entered the market with 2 million tons and thereafter, mini-mill

World and U. S. Markets for Steel and Steelmaking Raw Materials

Table 3.10. U.S. Steel Industry Capacity, Crude Steel Production, International Trade, Apparent Steel Consumption, Net Shipments, and Yield--1956-1987 (million metric tons except as noted)

Year	Gross Capacity	Crude Steel Production	Imports	Exports	Apparent Consumption	Net Shipments	Yield (percent)
1956	n.a.	104.49	1.18	3.90	101.77	75.55	65.6
1957	n.a.	102.22	1.09	4.90	98.41	72.47	64.3
1958	n.a.	77.91	1.54	2.54	76.91	54.33	63.3
1959	n.a.	84.71	3.99	1.54	87.16	62.95	67.4
1960	129.52	90.07	3.08	2.72	90.43	64.49	64.9
1961	n.a.	88.89	2.90	1.81	89.97	59.95	61.2
1962	n.a.	89.16	3.72	1.81	91.06	64.03	65.1
1963	n.a.	99.14	4.99	2.00	102.13	68.57	62.7
1964	n.a.	115.28	5.80	3.08	118.00	77.00	60.6
1965	134.42	119.27	9.43	2.27	126.44	84.08	63.9
1966	n.a.	121.63	9.80	1.54	129.88	81.63	60.9
1967	n.a.	115.37	10.43	1.54	124.26	76.10	59.8
1968	n.a.	119.27	16.33	2.00	133.60	83.35	63.4
1969	n.a.	128.16	12.70	4.72	136.14	85.17	60.3
1970	138.95	119.27	12.15	6.35	125.08	82.36	62.6
1971	n.a.	109.20	16.60	2.54	123.26	78.91	65.5
1972	n.a.	120.81	16.05	2.63	134.24	83.26	62.5
1973	141.00	136.78	13.79	3.72	146.84	101.04	67.0
1974	140.00	132.15	14.51	5.26	141.40	99.32	68.2
1975	138.86	105.76	10.88	2.63	114.01	72.56	62.2
1976	143.58	116.10	12.97	2.36	126.71	81.09	63.4
1977	145.12	113.65	17.51	1.81	129.34	82.63	65.9
1978	143.22	124.26	19.14	2.18	141.22	88.80	64.8
1979	140.86	123.62	15.87	2.54	136.96	90.97	66.7
1980	139.41	101.40	14.06	3.72	111.74	76.10	68.1
1981	139.95	109.57	18.05	2.63	124.98	80.27	66.5
1982	139.68	67.66	15.15	1.63	81.18	55.87	74.9
1983	136.59	76.73	15.51	1.09	91.15	61.31	72.5
1984	122.72	83.90	23.76	0.91	106.75	66.85	72.3
1985	121.18	80.09	22.04	0.82	101.31	66.21	75.0
1986	115.19	73.70	18.77	0.87	91.60	63.76	86.5
1987	101.77	80.91	18.50	1.00	98.42	69.57	86.0

Source: Compiled from American Iron and Steel Institute, *Annual Statistical Report*, Various Years.

capacity rose. In 1965, steel imports reached 9 million tons and peaked in 1971 at 17 million tons. After 1971, imports dropped to 11 million tons in 1975, followed by an increase to 18 million tons in 1986.

Thus, U.S. steel imports rose despite the protectionism discussed in Chapter 7. U.S. exports fell from 6 million tons in 1970 to 1 million tons in 1987. The low value of the dollar in 1987 and 1988 is expected to increase U.S. exports and reduce U.S. imports, which may improve U.S. industry operating rates.

Table 3.11. U.S. Major Mill Actual and "Break-even" Operating Rates by Quarters--1981 to 1986 (percent)

Year	Quarter	Operating Rates	
		Actual	Breakeven
1981	Q1	84	55
	Q2	85	42
	Q3	75	34
	Q4	62	60
1982	Q1	57	97
	Q2	47	115
	Q3	42	125
	Q4	36	124
1983	Q1	52	103
	Q2	60	90
	Q3	57	85
	Q4	59	75
1984	Q1	74	78
	Q2	77	67
	Q3	58	64
	Q4	54	70
1985	Q1	67	78
	Q2	70	67
	Q3	64	70
	Q4	63	72
1986	Q1	72	80
	Q2	68	70
	Q3	50	65
	Q4	72	77

Note:"Break-even" is the capacity utilization rate at which costs are recovered given prices and cost in each period.
Source: *World Steel Dynamics*, "Steel Survival Strategies II," 1987.

Table 3.14 shows the shipments of the U.S. steel industry by product. The flat-rolled products shown are plate, cold-rolled sheet, hot-rolled sheets, galvanized sheet, tin-mill products, other metallic and electrical sheets, and pipe and tubing (which are produced from plates). The non-flat-rolled products are structurals, hot-rolled bar, rod, and rail. From 1973 to 1987, the share of flat-rolled products shipments in the United States declined from 68 percent to 65 percent, while non-flat-rolled product shipments rose from 28 percent to 31 percent. Galvanized sheet was the only flat-rolled product with expanded shipments, rising from 6 millions to 8 million tons. Wire-rod shipments grew from 2 million to 6 million tons and was the only non-flat-rolled product with higher shipments.

Shipments of other flat-rolled and non-flat-rolled products also declined. Pipe and tubing, rails and accessories, wire-products, and plate shipment reduction was greater than 60 percent; tin-mill products and hot-rolled bar shipment reduction exceeded 40 percent; cold-rolled sheet and strip and

36   World and U. S. Markets for Steel and Steelmaking Raw Materials

Table 3.12. U.S. Net Manufacturers' Shipments, Exports, Imports, and Apparent Consumption of Steel by Steel Grade and U.S. Export and Import Prices--1973, 1979, 1980, 1983, 1986, and 1987

	1973	1979	1980	1983	1986	1987
Net Manufacturers' Shipments [a] (million metric tons)						
Carbon Steel	89.32	79.77	67.68	56.95	62.35	66.80
Alloy Steel	8.45	9.67	8.30	5.53	5.50	6.33
Stainless Steel	1.05	1.26	1.13	1.16	1.21	1.41
Exports (million metric tons)						
Carbon Steel	3.33	2.08	2.79	0.77	0.63	0.71
Alloy Steel	0.26	0.41	0.82	0.28	0.18	0.25
Stainless Steel	0.09	0.08	0.12	0.04	0.04	0.07
Imports (million metric tons)						
Carbon Steel	11.92	15.31	13.64	14.87	17.96	17.64
Alloy Steel	0.31	0.63	0.50	0.48	0.74	0.97
Stainless Steel	0.12	0.15	0.14	0.17	0.29	0.30
Apparent Consumption (million metric tons)						
Carbon Steel	97.92	93.00	78.53	71.05	79.69	83.74
Alloy Steel	8.50	9.89	7.98	5.73	6.06	7.05
Stainless Steel	1.08	1.34	1.15	1.29	1.45	1.64
Export Prices (dollars per metric ton)						
Carbon Steel	239.79	637.55	624.12	840.79	746.10	713.24
Alloy Steel	406.37	966.63	727.01	1084.18	1217.07	1090.45
Stainless Steel	1162.89	2287.82	2070.94	2598.78	2111.77	1819.31
Import Prices [b] (dollars per metric ton)						
Carbon Steel	192.95	475.45	525.36	454.32	433.64	457.71
Alloy Steel	465.48	946.58	1143.02	1103.81	862.85	888.01
Stainless Steel	1151.07	2459.78	2954.66	2276.29	1933.13	1949.92

Notes: [a] Includes quantities of steel consumed in steel producing plants in the manufacture of fabricated products and as maintenance, repair and operating supplies.
[b] U.S. import prices prior to 1976 were customs values, after 1976 tariffs, insurance and freight are included. Therefore, prices from 1973 to 1976 are not comparable with prices from 1977 to 1987.
Source: Compiled from Current Industrial Reports, Steel Mill Products, MA33B, U.S. Department of Commerce, Various Years.

structural fell 35 percent; hot-rolled sheet and strip went down by 28 percent; other bar products dropped 12 percent; other metallic and electrical sheets fell 6 percent. In 1987, despite a decrease in shipments, cold- and hot-rolled sheet and strip still had the largest share of U.S. steel shipments.

Another factor contributing to lower crude-steel production in the United States was the effect of indirect steel imports on steel-containing goods such as automobiles. In 1987, the U.S. net trade of steel-containing goods accounted for about 9 million tons of net imports, while imports and exports of steel-containing goods were almost equivalent in 1981 (Table 3.15). Net imports of steel-containing goods are expected to decline because foreign

Table 3.13. Major U.S. Steel Producers and Their Product Lines in 1981 and 1985 (percent of total shipments)

Product Year	USS	LTV	Beth-lehem	Inland	Armco	National
Sheet,Strip,Tin						
1985	66	63	42	69	77	100
1981	45	67	32	66	53	100
Plate,Structural						
1985	16	0	17	17	1	0
1981	23	0	21	19	14	0
Bar, Rod						
1985	7	31	8	14	17	0
1981	11	12	10	13	21	0
Pipe,Tubing						
1985	10	6	n.a.	0	4	0
1981	16	21	n.a.	0	8	0
Other						
1985	1	n.a.	3	0	1	0
1981	4	n.a.	8	2	4	0

Source: Chase Econometrics, 1987.

producers are undertaking production of their products in the United States and because of the dollar devaluation against strong currencies.

EFFECTS ON THE PRODUCTION AND TRADE OF IRON ORE, SCRAP AND SEMI-FINISHED PRODUCTS

Chapter 2 reviewed the alternative technologies available for steelmaking and the different ways in which steps in steelmaking could be located. The remainder of this chapter discusses developments in iron ore, scrap, pig iron, DRI, and semi-finished steel. Changes in both the final markets for steel and in the supply of the other commodities have affected the developments. In the scrap case, its supply was profoundly affected by the changes in steel technology reviewed in Chapter 2.

The coverage starts with 1977--two years before world output peaked. Thus, trends over the period shown are the net impact of a few years of growth and several years of decline. Overall, the trend was for increased reliance on scrap and lesser use of iron ore-derived inputs to steelmaking. Most of the low-cost producers of pig iron and merchant DRI are located offshore from the big market countries and the other noncommunist countries.

Table 3.14. Net Shipments of U.S. Steel Mills by Major Products Categories, All Grades--1973, 1979, 1980, 1983, 1986, and 1987 (million metric tons)

	1973	1979	1980	1983	1986	1987
Ingot, Bloom, Billet & Slab	3.36	2.36	2.39	0.91	1.35	1.44
Plates	8.78	8.19	7.33	3.46	3.23	3.67
Tin Mill Products	6.64	5.72	5.18	3.91	3.45	3.62
Hot-Rolled Sheet & Strip	17.01	15.40	11.60	11.11	11.61	12.43
Cold-Rolled Sheet & Strip	19.88	16.79	12.92	13.33	12.85	12.68
Galvanized Sheet	6.25	5.75	4.76	5.84	7.37	8.25
Other Metal & Electric Sheets	1.64	1.52	1.20	1.27	1.44	1.54
Pipe & Tubing	8.28	7.48	8.25	2.94	2.58	3.24
Structurals	5.95	4.81	4.41	3.05	3.84	4.39
Wire Rods	1.85	2.59	2.42	2.62	3.14	3.48
Wire Products	2.94	2.22	1.60	1.23	0.98	1.00
Hot-Rolled Bars	8.83	7.70	5.41	4.65	5.12	5.49
Other Bar Products	7.66	8.26	6.62	5.96	5.92	6.77
Rail & Accessories	1.17	1.48	1.31	0.74	0.51	0.39
Other Steel Products	0.83	0.66	0.67	0.28	0.34	1.14
Total Shipments	101.07	90.94	76.05	61.30	63.73	69.53

Source: Compiled from American Iron and Steel Institute, *Annual Statistical Report,* Various Years.

Iron Ore

About 98 percent of iron-ore production goes to blast furnaces to produce pig iron. As the prior discussion should have suggested, slower growth in steel output and shifts to scrap-based steel production have lessened iron-ore demand growth. World iron-ore production was 872 million tons in 1977, peaked at 931 million tons in 1979, and was 909 million tons in 1986. The respective international trade numbers were 355 million, 396 million, and 366 million tons.

Table 3.16 shows the production, imports, and exports of iron ore for the six country categories into which the noncommunist world was subdivided. The big market countries (United States, EEC, Japan) are net importers of iron ore. Japan and the United States lowered their net imports. From 1977 to 1986, the import penetration (the ratio of iron imports to apparent consumption) of iron ore in the big market countries rose from 75 percent to 84 percent because Japan and the EEC relied more on iron-ore imports. U.S. reliance on iron-ore imports fell from 42 percent to 33 percent.

From 1977 to 1986, the iron-rich countries expanded their net exports of iron ore from 234 million to 255 million tons. The iron-rich industrialized countries--Canada, Sweden, and Australia--dropped their net exports of iron ore from 140 million to 122 million tons. Iron-rich developing

Table 3.15. U.S. Net Trade in Steel-Containing Goods (million metric tons of steel content in the product)

Year	Exports	Imports	Net Exports
1979	9.2	10.4	-1.2
1980	9.1	9.6	-0.5
1981	9.5	9.7	-0.2
1982	8.3	9.6	-1.3
1983	7.2	10.5	-3.3
1984	8.8	13.8	-5.0
1985	8.7	16.1	-7.4
1986	8.2	17.5	-9.3
1987	8.3	16.8	-8.5

Source: *World Steel Dynamics*, "Steel Strategist #14," 1987.

countries increased their net exports from 94 million to 133 million tons; their iron-ore production grew from 150 million to 205 million tons. The rest of the iron-ore production rise was absorbed in higher production of pig iron, DRI, or steel.

Oil-rich developing countries lowered net exports of iron ore; this was especially true after 1980. Net exports of oil-rich countries with reserves of iron ore--Venezuela, Argentina, Mexico, and Algeria--fell from 12 million to 7 million tons, but their iron-ore production rose from 23 million to 29 million tons. The other oil-rich countries necessarily depended entirely on ore imports that went from 0 to 3 million tons.

The other noncommunist countries category includes Liberia and Mauritania, iron-ore exporters with no significant crude-steel production. In 1986, exports of iron ore from these two countries represented 63 percent of total iron-ore exports in this category. From 1977 to 1986, net exports of iron ore for the two countries combined declined from 26 million to 24 million tons. The rest of the other-noncommunist-countries group raised net imports of iron ore.

Most of the growth in production of iron ore among noncommunist countries took place in the iron-rich developing countries, the only group of nations to increase its iron-ore exports. With the notable exception of oil-rich developing countries having iron-ore reserves, all the other groups of nations decreased their iron-ore production.

Pig Iron and DRI

As the discussion in Chapter 2 implies, the price of scrap relative to pig iron and DRI and the quality of available scrap determine the proportion in which these products are used in steelmaking. The proportion of scrap use will vary inversely with its relative price and directly with the average scrap quality.

This section shows the role of pig iron and DRI for individual categories of countries. International trade of pig iron did not change significantly, and

Table 3.16. Apparent Consumption, Production, and International Trade of Iron Ore by Regions[a] of the World--1977, 1979, 1980, 1983, and 1986

	1977	1979	1980	1983	1986
Production of Iron Ore (million metric tons)					
U.S.	56.28	86.49	70.73	38.57	39.61
E.E.C.	54.93	49.15	43.11	24.81	19.67
Japan	0.68	0.46	0.48	0.30	0.29
Big Market	111.89	136.10	114.32	63.69	59.58
Iron-Rich Industrial	175.77	165.66	172.88	114.50	149.48
Iron-Rich Developing	150.02	166.57	163.66	146.29	205.30
Total Iron-Rich	325.79	332.22	336.54	260.79	354.78
Oil-Rich Dev. with Iron Ore	22.46	24.90	28.84	21.77	29.38
Other Western	61.73	61.68	62.04	52.18	58.58
Total Western	521.87	554.91	541.74	398.43	502.31
C.P.E.	349.68	376.09	372.04	372.81	407.05
Total World	871.55	931.00	913.78	771.24	909.36
Exports of Iron Ore (million metric tons)					
U.S.	2.18	5.23	5.78	3.84	4.55
E.E.C.	13.62	12.30	11.09	6.24	6.35
Japan	0.00	0.00	0.00	0.00	0.00
Big Market	15.79	17.53	16.87	10.08	10.90
Iron-Rich Industrial	142.85	153.47	140.37	113.90	127.67
Iron-Rich Developing	93.98	113.66	119.04	99.77	133.36
Total Iron-Rich	236.83	267.13	259.41	213.67	261.03
Oil-Rich Dev. with Iron Ore	12.94	15.46	13.18	7.55	10.08
Other Western	48.13	51.18	48.80	37.56	37.18
Total Western	313.69	351.30	338.26	268.86	319.18
C.P.E.	40.95	44.50	46.87	42.01	46.95
Total World	354.64	395.79	385.13	311.66	366.13
Iron Ore Imports (million metric tons)					
U.S.	38.52	34.35	25.49	13.54	16.97
E.E.C.	119.59	141.64	129.02	99.66	118.22
Japan	132.57	130.27	133.72	109.15	115.23
Big Market	290.68	306.25	288.24	222.36	250.42
Iron-Rich Industrial	2.63	5.95	5.88	4.03	5.79
Oil-Rich Dev. with Iron Ore	0.80	2.00	2.00	1.67	2.90
Other Oil-Rich Dev.	0.00	0.50	1.00	1.23	2.90
Total Oil-Rich Dev.	0.80	2.50	3.00	2.90	5.80
Other Western	11.98	20.91	23.83	23.28	27.30
Total Western	306.09	335.61	320.94	252.57	289.31
C.P.E.	54.94	60.53	62.48	51.15	66.02
Total World	361.03	396.14	383.43	303.72	355.33
Iron Ore Apparent Consumption (million metric tons)					
U.S.	92.62	115.61	90.44	48.28	52.03
E.E.C.	160.90	178.49	161.04	118.24	131.54
Japan	133.26	130.73	134.20	109.45	115.53

Table 3.16. (continued)

	1977	1979	1980	1983	1986
Big Market	386.78	424.83	385.68	275.97	299.09
Iron-Rich Industrial	35.55	18.13	38.39	4.64	27.60
Iron-Rich Developing	56.04	52.91	44.62	46.52	71.94
Total Iron-Rich	91.58	71.04	83.01	51.16	99.54
Oil-Rich Dev.with Iron Ore	10.32	11.44	17.65	15.89	22.20
Other Oil-Rich Dev.	0.00	0.50	1.00	1.23	2.90
Total Oil-Rich Dev.	10.32	11.94	18.65	17.12	25.10
Other Western	25.59	31.41	37.08	37.90	48.70
Total Western	514.26	539.22	524.42	382.15	472.43
C.P.E.	363.67	392.12	387.65	381.95	426.13
Total World	877.94	931.35	912.07	763.30	898.56
Iron Ore Net Exports (million metric tons)					
U.S.	-36.34	-29.12	-19.71	-9.70	-12.41
E.E.C.	-105.97	-129.33	-117.93	-93.43	-111.87
Japan	-132.57	-130.27	-133.72	-109.15	-115.23
Big Market	-274.89	-288.72	-271.36	-212.28	-239.52
Iron-Rich Industrial	140.22	147.53	134.49	109.86	121.88
Iron-Rich Developing	93.98	113.66	119.04	99.77	133.36
Total Iron-Rich	234.21	261.18	253.53	209.64	255.24
Oil-Rich Dev.with Iron Ore	12.14	13.46	11.18	5.88	7.18
Other Oil-Rich Dev.	0.00	-0.50	-1.00	-1.23	-2.90
Total Oil-Rich Dev.	12.14	12.96	10.18	4.65	4.28
Other Western	36.14	30.27	24.97	14.28	9.88
Total Western	7.60	15.69	17.32	16.29	29.88
C.P.E.	-13.99	-16.03	-15.61	-9.14	-19.07
Total World[b]	-6.39	-0.35	1.71	7.95	10.80

Notes: [a] See text for definition of regions.
[b] Net represents discrepancies between import and export statistics.
Source: International Iron and Steel Institute, *Steel Statistics Yearbook*, Various Years.

most merchant DRI production was consumed domestically. From 1977 to 1986, pig-iron international trade averaged 10 million tons per year, but each country category behaved differently (Table 3.17).

From 1977 to 1986, the big market countries lowered pig-iron production from 254 million to 199 million tons, and their net imports of pig iron were maintained at an average of 1 million tons per year. Net pig-iron exports from iron-rich industrialized countries diminished from 1.4 to 0.6 million. The iron-rich developing countries raised their net exports of pig iron from 1.6 million to 2.2 million tons. Therefore, some of the growth of iron-ore production in the iron-rich developing countries was directed toward exporting pig iron mainly to the other-noncommunist-country group.

From 1977 to 1986, the oil-rich developing countries reduced their net imports of pig iron, and in 1986 they became net exporters of pig iron,

Table 3.17. Apparent Consumption, Production, and International Trade of Pig Iron by Regions[a] of the World--1977, 1979, 1980, 1983, and 1986

	1977	1979	1980	1983	1986
Production of Pig Iron (million metric tons)					
U.S.	73.78	78.93	62.34	44.21	39.77
E.E.C.	93.90	104.51	95.55	79.58	84.54
Japan	85.89	83.83	87.04	72.94	74.65
Big Market	253.56	267.27	244.93	196.73	198.96
Iron-Rich Industrial	18.72	21.57	20.25	15.64	17.54
Iron-Rich Developing	24.99	27.38	28.41	27.32	36.92
Total Iron-Rich	43.71	48.95	48.66	42.96	54.45
Oil-Rich Dev.with Iron Ore	4.83	5.60	5.83	5.73	6.66
Other Oil-Rich Dev.	0.92	0.95	1.23	1.14	1.20
Total Oil-Rich Dev.	5.75	6.55	7.06	6.87	7.86
Other Western	13.62	19.89	20.20	24.47	27.40
Total Western	316.65	342.66	320.85	271.02	288.67
C.P.E.	169.12	185.62	187.21	186.24	207.43
Total World	485.77	528.28	508.05	457.27	496.10
Exports of Pig Iron (million metric tons)					
U.S.	0.06	0.11	0.09	0.02	0.05
E.E.C.	1.09	1.57	1.50	0.80	0.78
Japan	0.58	0.06	0.02	0.35	1.07
Big Market	1.73	1.75	1.61	1.16	1.91
Iron-Rich Indust. Countries	1.52	1.17	1.26	0.83	0.70
Iron-Rich Dev. Countries	1.55	1.15	0.96	1.80	2.37
Total Iron-Rich Countries	3.08	2.32	2.22	2.63	3.07
Oil-Rich Dev.Countries with Iron Ore	0.00	0.00	0.00	0.00	0.00
Other Oil-Rich Dev. Countries	0.00	0.00	0.00	0.06	0.09
Total Oil-Rich Dev. Countries	0.00	0.00	0.00	0.06	0.09
Other Western	0.28	0.29	0.26	0.05	0.05
Total Western	5.08	4.35	4.09	3.91	5.12
C.P.E.	4.72	4.54	4.56	4.55	4.56
Total World	9.80	8.89	8.65	8.46	9.68
Imports of Pig Iron (million metric tons)					
U.S.	0.47	0.53	0.42	0.34	0.36
E.E.C.	2.06	1.92	2.04	1.32	1.75
Japan	0.55	0.56	0.79	1.00	0.97
Big Market	3.08	3.01	3.25	2.66	3.08
Iron-Rich Industrial	0.10	0.07	0.06	0.05	0.08
Iron-Rich Developing	0.00	0.02	0.00	0.20	0.20
Total Iron-Rich	0.10	0.09	0.06	0.26	0.28
Oil-Rich Dev.with Iron Ore	0.14	0.05	0.15	0.21	0.00
Other Oil-Rich Dev.	0.02	0.03	0.04	0.09	0.08
Total Oil-Rich Dev.	0.15	0.08	0.19	0.29	0.08
Other Western	0.44	0.63	0.38	0.68	1.09
Total Western	3.77	3.82	3.88	3.89	4.53
C.P.E.	5.93	4.66	4.59	3.72	4.19
Total World	9.70	8.48	8.47	7.61	8.72
Apparent Consumption of Pig Iron (million metric tons)					
U.S.	74.18	79.34	62.68	44.53	40.08
E.E.C.	94.87	104.86	96.08	80.10	85.51

Table 3.17. (continued)

	1977	1979	1980	1983	1986
Japan	85.86	84.33	87.82	73.59	74.55
Big Market	254.91	268.53	246.57	198.23	200.13
Iron-Rich Industrial	17.30	20.47	19.05	14.86	16.91
Iron-Rich Developing	23.44	26.25	27.45	25.73	34.75
Total Iron-Rich	40.74	46.72	46.50	40.59	51.66
Oil-Rich Dev. with Iron Ore	4.97	5.65	5.99	5.93	6.66
Other Oil-Rich Dev.	0.93	0.98	1.26	1.16	1.19
Total Oil-Rich Dev.	5.90	6.63	7.25	7.10	7.85
Other Western	13.79	20.23	20.32	25.09	28.44
Total Western	315.34	342.12	320.64	271.01	288.08
C.P.E.	170.33	185.74	187.24	185.41	207.06
Total World	485.66	527.86	507.87	456.41	495.14
Net Exports of Pig Iron (million metric tons)					
U.S.	-0.40	-0.41	-0.33	-0.32	-0.30
E.E.C.	-0.97	-0.35	-0.53	-0.52	-0.98
Japan	0.03	-0.50	-0.78	-0.66	0.11
Big Market	-1.35	-1.26	-1.64	-1.50	-1.17
Iron-Rich Industrial	1.42	1.10	1.20	0.78	0.63
Iron-Rich Developing	1.55	1.13	0.96	1.60	2.17
Total Iron-Rich	2.98	2.23	2.16	2.37	2.80
Oil-Rich Dev. with Iron Ore	-0.14	-0.05	-0.15	-0.21	0.00
Other Oil-Rich Dev.	-0.02	-0.03	-0.04	-0.03	0.01
Total Oil-Rich Dev.	-0.15	-0.08	-0.19	-0.23	0.01
Other Western	-0.16	-0.35	-0.12	-0.63	-1.04
Total Western	1.31	0.54	0.21	0.02	0.59
C.P.E.	-1.21	-0.12	-0.03	0.84	0.37
Total World[b]	0.10	0.42	0.18	0.85	0.96

Notes: [a] See text for definition of regions.
[b] Net represents discrepancies between import and export statistics.
Source: International Iron and Steel Institute, *Steel Statistics Yearbook*, Various Years.

albeit by the very insignificant amount of 10,000 tons. Other noncommunist countries expanded net imports of pig iron from 0.2 million to 1 million tons.

World production of DRI rose from 3 million to 13 million tons. Most of this growth was in the oil-rich countries. Their output grew from 2 million to 9 million tons. In 1986, these countries accounted for 72 percent of world DRI production. From 1977 to 1986, the iron-rich countries expanded production of DRI from 1 million to 2 million tons; big market countries lowered DRI production from 0.7 million to 0.3 million tons (Table 3.18).

DRI capacity went up in developing countries. The cost of merchant DRI production in these developing countries is half that of DRI costs in the United States. Although the operating rate averaged 59 percent in 1984, those operating rates differed from country to country. Qatar, Saudi Arabia, and Argentina had operating rates above 91 percent, while plants in Iraq and

Table 3.18. Direct Reduction Production by Regions[a] of the World--1977, 1979, 1980, 1983, and 1986

	1977	1979	1980	1983	1986
Direct Reduction Production (million metric tons)					
U.S.	0.52	0.81	0.65	0.00	0.16
E.E.C.	0.20	0.48	0.42	0.07	0.17
Japan	0.00	0.00	0.00	0.00	0.00
Big Market	0.72	1.29	1.07	0.07	0.33
Iron-Rich Industrial	0.36	0.91	0.88	0.54	0.69
Iron-Rich Developing	0.36	0.43	0.40	0.33	1.21
Total Iron-Rich	0.72	1.34	1.28	0.87	1.90
Oil-Rich Dev. with Iron Ore	1.75	3.38	4.03	4.83	5.20
Other Oil-Rich Dev.	0.00	0.69	0.79	1.71	3.53
Total Oil-Rich Dev.	1.75	4.07	4.82	6.53	8.73
Other Western	0.12	0.16	0.18	0.20	0.90
Total Western	3.31	6.86	7.35	7.67	11.85
C.P.E.	0.00	0.00	0.00	0.02	0.75
Total World	3.31	6.86	7.35	7.69	12.60

Note: [a] See text for definition of regions.

Source: International Iron and Steel Institute, *Steel Statistics Yearbook*, Various Years.

Iran were inactive. In Brazil, Peru, and Mexico, operating rates exceeded the 1984 average, but the operating rates in Venezuela, Indonesia, Nigeria, Trinidad, South Africa, and Malaysia were below the 1984 average (Innace, 1985a).

DRI capacity in the United States increased modestly until 1982, but since then has declined (*World Steel Dynamics*, 1987). As of early 1989, the Georgetown Steel Corporation in South Carolina was the only domestic steel producer operating a DRI plant. Two other producers of merchant DRI, Armco and Oregon Steel, have ceased production in their DRI plants (*World Steel Dynamics*, 1987). The same trend towards reduced DRI production arose in industrialized countries. DRI capacity utilization in the industrialized countries was 29 percent in 1986.

Therefore, the iron-rich developing countries and the oil-rich developing countries expanded their production of pig iron and DRI to raise their domestic production of crude steel, and, in the case of iron-rich developing countries, to supply the international pig-iron market. Other noncommunist countries increased pig-iron imports. The big market and industrialized iron-rich countries reduced pig iron and DRI production.

Scrap

Table 3.19 shows the apparent production, imports, exports, net exports, and consumption of scrap for each of the countries' categories. The big

Table 3.19. Consumption, Apparent Production, and International Trade of Scrap by Regions[a] of the World--1977, 1979, 1980, 1983, and 1986

	1977	1979	1980	1983	1986
Apparent Production of Scrap (million metric tons)					
U.S.	67.93	79.46	70.01	50.69	54.86
E.E.C.	71.29	73.48	68.72	64.51	61.87
Japan	33.38	42.43	40.98	36.37	37.50
Big Market	172.60	195.37	179.71	151.57	154.22
Iron-Rich Industrial	10.11	11.51	11.51	9.61	9.73
Iron-Rich Developing	0.31	0.00	0.00	0.00	0.00
Total Iron-Rich	10.41	11.51	11.51	9.61	9.73
Other Western	4.05	4.91	4.69	4.70	5.21
Total Western	187.06	211.78	195.91	165.88	169.16
C.P.E.	80.86	83.34	84.16	83.73	85.34
Total World	265.16	291.08	275.60	244.53	248.69
Exports of Scrap (million metric tons)					
U.S.	5.60	10.13	10.13	6.82	10.67
E.E.C.	8.29	9.74	11.12	12.43	13.31
Japan	0.21	0.15	0.16	0.12	0.46
Big Market	14.10	20.01	21.41	19.37	24.44
Iron-Rich Industrial	1.32	1.63	1.51	1.48	0.87
Iron-Rich Developing	0.21	0.01	0.01	0.01	0.01
Total Iron-Rich	1.53	1.63	1.51	1.49	0.87
Oil-Rich Dev. Countries with Iron Ore	0.00	0.00	0.00	0.00	0.00
Other Oil-Rich Dev. Countries	0.00	0.00	0.00	0.00	0.03
Total Oil-Rich Dev. Countries	0.00	0.00	0.00	0.00	0.03
Other Western	0.41	0.55	0.54	1.37	1.06
Total Western	16.03	22.20	23.47	22.23	26.40
C.P.E.	1.95	2.17	2.60	3.47	3.32
Total World	17.98	24.36	26.17	25.69	29.71
Imports of Scrap (million metric tons)					
U.S.	0.57	0.69	0.51	0.58	0.66
E.E.C.	10.45	14.19	15.41	12.98	13.74
Japan	1.44	3.35	2.99	3.91	3.22
Big Market	12.46	18.22	18.91	17.47	17.62
Iron-Rich Industrial	0.62	1.18	1.09	1.12	1.09
Iron-Rich Developing	0.06	0.08	0.10	0.80	1.00
Total Iron-Rich	0.67	1.26	1.19	1.92	2.09
Oil-Rich Dev. Countries with Iron Ore	0.36	0.74	1.05	0.38	0.46
Other Oil-Rich Dev. Countries	0.05	0.03	0.04	0.26	0.19
Total Oil-Rich Dev. Countries	0.40	0.77	1.09	0.64	0.65
Other Western	3.71	4.67	4.94	6.95	8.70
Total Western	17.24	24.92	26.13	26.98	29.06
C.P.E.	0.53	0.73	1.20	0.71	1.00
Total World	17.78	25.65	27.32	27.69	30.06
Consumption of Scrap (million metric tons)					
U.S.	62.90	70.03	60.38	44.45	44.85
E.E.C.	73.46	77.93	73.01	65.06	62.30
Japan	34.61	45.62	43.81	40.16	40.26
Big Market	170.96	193.58	177.20	149.67	147.40
Iron-Rich Industrial	9.40	11.06	11.10	9.24	9.95

Table 3.19. (continued)

	1977	1979	1980	1983	1986
Iron-Rich Developing	0.15	0.08	0.09	0.80	1.00
Total Iron-Rich	9.55	11.13	11.19	10.04	10.95
Oil-Rich Dev.Countries with Iron Ore	0.36	0.74	1.05	0.38	0.46
Other Oil-Rich Dev. Countries	0.05	0.03	0.04	0.26	0.16
Total Oil-Rich Dev. Countries	0.40	0.77	1.09	0.64	0.62
Other Western	7.35	9.03	9.09	10.28	12.85
Total Western	188.27	214.51	198.57	170.63	171.82
C.P.E.	79.44	81.90	82.76	80.97	83.03
Total World	264.95	292.37	276.76	246.52	249.04
Net Exports of Scrap (million metric tons)					
U.S.	5.04	9.44	9.63	6.24	10.02
E.E.C.	-2.17	-4.45	-4.29	-0.55	-0.43
Japan	-1.23	-3.20	-2.83	-3.79	-2.76
Big Market	1.64	1.79	2.51	1.90	6.82
Iron-Rich Industrial	0.71	0.45	0.42	0.37	-0.23
Iron-Rich Developing	0.15	-0.08	-0.09	-0.80	-1.00
Total Iron-Rich	0.86	0.37	0.32	-0.43	-1.22
Oil-Rich Dev.Countries with Iron Ore	-0.36	-0.74	-1.05	-0.38	-0.46
Other Oil-Rich Dev. Countries	-0.05	-0.03	-0.04	-0.26	-0.16
Total Oil-Rich Dev. Countries	-0.40	-0.77	-1.09	-0.64	-0.62
Other Western	-3.31	-4.12	-4.40	-5.58	-7.64
Total Western	-1.21	-2.73	-2.66	-4.76	-2.66
C.P.E.	1.41	1.44	1.40	2.76	2.32
Total World[b]	0.21	-1.29	-1.16	-1.99	-0.35

Notes: [a] See text for definition of regions.
[b] Net represents discrepancies between import and export statistics.
Source: International Iron and Steel Institute, *Steel Statistics Yearbook*, Various Years.

market countries are the largest producers of scrap. In particular, the United States is the largest producer and net exporter of scrap and the only noncommunist country that is a net exporter of scrap.

From 1977 to 1986, international scrap trade grew from 18 million to 30 million tons. Most of this international trade was among big market countries. In 1986, these big market countries, mainly the United States and the EEC, exported 80 percent of the scrap traded internationally and imported 60 percent of the scrap. From 1977 to 1986, the net scrap exports of big market countries went from 2 million to 7 million tons. During this period, U.S. net exports of scrap doubled from 5 million to 10 million tons. The remaining two big market countries are net importers of scrap, although the EEC decreased its scrap net imports.

The iron-rich countries were net exporters of scrap until 1980 and then became net importers. Since 1981, the iron-rich industrialized countries raised their dependence on scrap imports. Iron-rich developing countries also boosted net imports of scrap. The developing countries do not generate large amounts of scrap. The iron-rich and oil-rich developing countries chose to base steel production on domestic production of pig iron or DRI instead of importing scrap. From 1977 to 1980, the oil-rich developing countries increased net imports of scrap. After 1981, they maintained their net imports of scrap while raising crude-steel production. Scrap imports are very suitable for these countries because their crude-steel production processes are based mostly on direct reduction, electric furnaces, and continuous casting. However, growth of DRI production made these countries less dependent on scrap imports. The other noncommunist countries increased their net imports of scrap from 3 million to 8 million tons. Therefore, the other noncommunist countries expanded their net imports of iron ore, pig iron, and scrap.

In the period analyzed, international scrap trade grew at an average annual rate of 1 million tons while pig-iron trade maintained its volume. Most of the scrap trade took place in the big market countries and, to a lesser extent, the iron-rich industrialized countries. The substitution of scrap for pig iron resulted from the economics of the input markets for the industrialized countries. The availability of scrap in the industrialized countries made greater substitution of scrap for pig iron more profitable (see Chapter 6 for further discussion).

Their reliance on scrap-charging in the steelmaking processes rose because scrap was readily available at relatively lower prices than other semi-finished iron products. As the following discussion of scrap supplies shows, this cost difference could close if world steel demand rose sharply enough. Steel consumers and steel producers, particularly in the big market countries, adopted technologies that generated less scrap and lowered the apparent world supply of scrap from 265 million to 249 million tons. The fall in the big market countries was from 173 million to 154 million tons.

Apparent U.S. production of scrap went from 68 million to 55 million tons. From 1973 to 1986, U.S. consumption of scrap fell from 92 million to 59 million tons, although scrap consumption for the EF rose from 26 million to 29 million tons. During this time, scrap consumption for the BOF and the OH process declined by 30 million to 15 million tons and from 19 million to 1 million tons, respectively. This represents a drop in the share of scrap charging from 40 percent to 34 percent in the BOF and 51 percent to 40 percent in the OH (*World Steel Dynamics*, 1987).

As the supply of scrap or any other commodity falls, its price is expected to rise, as was the case in 1987-88. Scrap played an important role in reducing the production costs of the big market countries, iron-rich industrialized countries, other oil-rich developing countries, and other noncommunist countries.

Semi-Finished Steel

Importing semi-finished steel is equivalent to substituting foreign for domestic production of crude steel. The attractiveness of such trade necessarily depends upon whether semi-finished steel can be obtained more cheaply by imports or through domestic production. Semi-finished steel historically was not a significant item in shipments to steel consumers and was traded almost exclusively among steel producers. Traditionally, steel producers entered the market to test another producer's materials or to cover a temporary shortfall in their own production capacity, such as when a blast furnace is relined (U.S. International Trade Commission, 1988).

As discussed further in Chapters 4, 5, and 6, U.S. steel industry reorganization and high capital costs for conventional technology altered the market for semi-finished steel. Over-capacity in the noncommunist world and the need to bring this in line with lower levels of demand eliminated interest in semi-finished steel imports in the 1970s to early 1980s. Subsequent industry evolution stimulated the emergence of the semi-finished steel market. Slab trade has risen and is expected to continue to increase. In particular, part of world flat-rolled production may be fabricated in consuming countries from imported slab.

The semi-finished steel market resulted from economic changes that increase the profitability of processing the iron ore prior to export. The low costs of inputs, such as labor in developing countries and natural gas in oil-rich countries, and the transferability of steel technology which allows low cost steel production, facilitated semi-finished steel exports.

Persistent demands for semi-finished steel are fundamentally based on differences in the cost of steelmaking. Joint ventures and long-run supply arrangements will be established. Finishing capacity will become greater than steelmaking capacity. Examples of plants where finishing capacity exceeds steelmaking capacity are Nasco (Philippines) and Eregli (Turkey). This type of demand also includes the plants projected only for re-rolling purposes, such as CSI and Tuscaloosa in the United States. Temporary demand results from accidents in the hot-end facilities, relining of blast furnaces and other equipment in the hot-end of steel production, expansion of the finishing capacity before the steelmaking capacity, and spot supply of finished steel products.

Growth in the semi-finished steel market predominantly involved persistent flows. The semi-finished steel market provides a way of satisfying future demand for steel by shifting steel expansion to areas in the world that have comparative advantages in steelmaking, such as the iron-ore reserves in iron-rich developing countries and some oil-rich developing countries.

The semi-finished steel market bridges traditional technology and new technology by permitting established steel producers to continue operations without making additional investments in steelmaking facilities. The tendency for scrap supply to decline, steel industry-restructuring, and development of new technology, semi-finished steel all contributed to the rise in the semi-finished steel trade. Reasons exist for believing that even more such trade would be profitable. The economic attractiveness of developing semi-finished steel production for export in these countries to some market countries is assessed in Chapter 6.

Bringing a low-value, high-volume material such as iron ore from low-cost producer countries is inefficient when low-cost semi-finished steel can be imported. Shipping a more processed product tends to produce transportation cost savings because less waste material must be shipped. Higher processing costs in the ore-producing country may offset the transportation cost advantage, but, as Chapter 6 seeks to demonstrate, production costs in such countries may be lower than in leading industrialized countries.

Steady supplies of semi-finished steel originate from the plants where the steelmaking capacity exceeds the finishing capacity, restriction of finishing products capacity (the EEC), restriction on the market (ISCOR), and international agreements (USS and Posco agreement). Examples of plants where steelmaking capacity exceeds finishing capacity are Tubarao (Brazil), SSAB (Sweden), and the Hoogovens (Netherlands). The Brazilian case is different from the other two because the plant was designed to produce semi-finished steel for export. Temporary supplies of semi-finished steel result from refurbishment, expansion, or accidents in the finishing capacity.

Steelmaking is rising in iron-ore and oil-rich developing countries and declining in big market countries. Some of these new producers may be low-cost sources of semi-finished steel for existing large steel plants in industrialized countries, such as the United States. Ecuador, Malaysia, Philippines, Nigeria, Greece, and Turkey--all in the "other noncommunist" group--are possible suppliers.

From 1977 to 1985, international trade in semi-finished steel grew from 8 million to 14 million tons when hot-rolled coil for re-rolling is not considered, and 11 million to 19 million tons when it is considered. In 1988, international trade of semi-finished steel shapes rose to an estimated level of 12 million tons, and slab trade to an estimated level of 8 million tons per year. Table 3.20 shows international trade of semi-finished steel from 1977 to 1985. International trade of semi-finished steel averaged 9 million tons per year until 1983, then increased to 14 million tons in 1985. The semi-finished steel share of world total steel imports averaged 6.3 percent from 1977 to 1983 and then grew to 8.5 percent in 1985 (Table 3.21). The expected range in the mid-1990s is for 10 million to 17 million tons (*World Steel Dynamics*, 1988).

The big market countries imported 3.5 million tons in 1977 and 6 million tons in 1985. The EEC accounted for about 3 million tons of imports throughout this period but expanded exports from 3 million to 6 million tons. U.S. imports rose from less than 0.4 million to 2.2 million tons. U.S. exports fell from about 230,000 tons to 80,000 tons. The United States was the largest importer of semi-finished steel. From 1984 to 1987, slab imports went from 0.7 million to 1.4 million tons, of which carbon steel grade accounts for 99 percent. The share of carbon-steel-slab imports from total semi-finished steel grew from 51 percent to 57 percent during the period 1984 to 1987. By July 1988, carbon-steel-slab imports reached 1.1 million tons (Table 3.22). Japanese imports were small but rising; exports, small but falling.

The iron-rich industrialized countries imported an average of 0.3 million tons of semi-finished steel per year throughout the period 1977-85. Exports from iron-rich industrialized countries moved irregularly with a tendency to decrease. The 1977 level was 2 million tons; the 1985 level, 710,000 tons.

The imports of iron-rich developing countries were maintained at very low levels from 1977 to 1985. Exports of iron-rich developing countries declined from about one million tons in 1977 to 260,000 tons in 1981.

Table 3.20. International Trade of Semi-Finished Steel by Regions[a] of the Noncommunist World--1977, 1979, 1980, 1983, and 1985

	1977	1979	1980	1983	1985
Imports of Semi-Finished Steel					
U.S.	0.38	0.37	0.13	0.96	2.22
E.E.C.	3.03	3.89	3.64	2.29	3.26
Japan	0.11	0.50	0.43	0.13	0.38
Big Market	3.52	4.76	4.20	3.38	5.86
Iron-Rich Industrial	0.13	0.25	0.23	0.14	0.24
Iron-Rich Developing	0.01	0.07	0.06	0.05	0.10
Total Iron-Rich	0.14	0.32	0.29	0.19	0.33
Oil-Rich Dev.Countries with Iron Ore	0.79	0.56	0.98	0.19	0.19
Other Oil-Rich Dev. Countries	0.37	0.54	0.69	0.88	0.69
Total Oil-Rich Dev. Countries	1.16	1.10	1.67	1.07	0.88
Other Western	1.99	2.41	2.33	3.23	4.38
Total Western	6.80	8.56	8.46	7.84	11.37
C.P.E.	0.30	0.35	0.24	0.22	1.97
Unallocated	1.04	0.95	0.17	0.57	0.16
Total World	8.14	9.86	8.88	8.63	13.50
Exports of Semi-Finished Steel					
U.S.	0.23	0.32	0.83	0.09	0.08
E.E.C.	2.95	3.65	3.76	3.71	5.82
Japan	0.43	1.23	0.19	0.13	0.32
Big Market	3.61	5.20	4.78	3.94	6.22
Iron-Rich Industrial[b]	2.06	1.52	0.73	1.18	0.71
Iron-Rich Developing[c]	0.94	0.69	0.43	0.66	2.43
Total Iron-Rich[bc]	3.00	2.22	1.16	1.84	3.14
Other Western[d]	0.50	1.43	1.58	1.83	0.59
Total Western	7.11	8.85	7.51	7.60	10.89
C.P.E.[e]	1.03	0.87	0.34	0.44	n.a
Unallocated	0.00	0.13	1.02	0.59	2.61
Total World	8.14	9.86	8.88	8.63	13.50
Net Exports of Semi-Finished Steel					
U.S.	-0.15	-0.04	0.70	-0.86	-2.14
E.E.C.	-0.07	-0.24	0.11	1.43	2.56
Japan	0.32	0.73	-0.24	0.00	-0.06
Big Market[b]	0.09	0.45	0.58	0.56	0.36
Iron-Rich Industrial	1.93	1.28	0.50	1.04	0.47
Iron-Rich Developing	0.94	0.62	0.37	0.61	2.33
Total Iron-Rich	2.87	1.90	0.87	1.65	2.81
Other Western	-1.49	-0.98	-0.75	-1.40	-3.79
Total Western	0.31	0.29	-0.95	-0.24	-0.47
C.P.E.	0.73	0.52	0.10	0.22	n.a
Unallocated	-1.04	-0.81	0.85	0.02	2.45
Total World	0.00	0.00	0.00	0.00	0.00

Notes:[a] See text for definition of regions.

[b] Exports of semi-finished for Australia are not available for 1985.

[c] Exports of semi-finished for India and South Africa are not available for 1985.

Subsequent rises brought the level to 2.4 million tons in 1985. Brazil accounts for at least 50 percent of these exports (see below).

In the 1977-86 period, imports of semi-finished steel by the oil-rich developing countries grew from 1977 to 1980 but then began to drop. Although the data on semi-finished exports for the oil-rich countries are not tabulated in the source employed here, the oil-rich countries were not important net exporters of semi-finished steel from 1977 to 1985.

Other noncommunist countries increased semi-finished steel imports from 2 million to 4 million tons. Exports of semi-finished steel averaged 1 million tons per year for which some Western European countries and South Korea were the main source. Rising dependence on semi-finished steel imports by other noncommunist countries resulted in a growth of net imports from 2 million to 4 million tons. Some countries, such as Turkey, the Philippines, Greece, and Malaysia, became more dependent on semi-finished steel imports.

A few other noncommunist countries were able to export some semi-finished steel, even though the overall group (excluding Mauritania and Liberia) was a net importer of semi-finished products and iron ore. South Korea was the primary supplier of semi-finished steel because of low labor costs, the adoption of modern technology, and deep water ports. These conditions made Korea one of the world's lowest cost steel producers.

The interaction between semi-finished steel imports and steel production is summarized in Table 3.23. The table shows that eighteen countries had shares of semi-finished steel imports in total steel imports of at least 8 percent in 1985. Of these eighteen countries, eight countries have semi-finished steel imports accounting for more than 20 percent of their crude-steel production (Table 3.23). Most of them play a major role in international trade of steel finished products.

Of the group of countries listed in Table 3.23, imports of semi-finished steel shapes for the United States, China, and Turkey in 1985 accounted for at least 10 percent of total semi-finished shapes traded internationally (Table 3.24). The Philippines and four industrialized nations--the United States, Italy, Belgium, and Japan--lowered crude-steel production. The other thirteen countries expanded their crude-steel production during the period analyzed.

The main semi-finished product exporting countries in 1987 are listed in Table 3.25. Brazil is the world's main supplier of semi-finished steel shapes.The other consistent main suppliers of steel slabs were Sweden, West Germany, the United Kingdom, and France. Other regular suppliers were the Netherlands, Belgium, Spain, South Korea, Canada, South

---

Table 3.20. (continued)

[d] Exports of semi-finished shapes for Taiwan are not available for 1985

[e] Exports of semi-finished shapes all CPE countries are not available for 1985.

Source: Iron and Steel Statistics Bureau, *World Steel Statistics Summary*, Various Years.

Table 3.21. Share of Semi-Finished Steel Imports for Selected Countries[a]--1977, 1979, 1980, 1983, and 1985 (percent)

Country	1977	1979	1980	1983	1985
Philippines	25	29	20	29	60
Turkey	11	11	14	23	45
Greece	9	28	22	31	34
Ecuador	22	20	40	26	30
Yugoslavia	18	18	16	1	22
Mexico	2	5	15	12	22
Spain	5	12	5	4	18
Bulgaria	3	0	0	0	17
Italy	7	12	13	7	16
Malaysia	0	12	9	29	16
Tunisia	31	8	0	5	14
Japan	50	34	37	5	14
Belgium	11	15	17	7	13
Nigeria	1	0	1	5	11
U.S.	2	2	1	6	10
Singapore	1	5	4	10	10
Austria	9	8	10	5	8
China	0	1	3	1	8

Note: [a] Includes all countries with a 1987 share of semi-finished steel in total steel imports of at least eight percent.

Source: International Iron and Steel Institute, *Steel Statistics Yearbook,* Various Years and Iron and Steel Statistics Bureau, *World Steel Statistics Summary,* Various Years.

Africa, Australia, and Japan. Mexico and Venezuela raised their shipments and may be among the future leading semi-finished steel exporters.

International trade of slab is very concentrated with Brazil (Tubarao plant) supplying 46 percent of world market needs in 1987, followed by the United Kingdom (British Steel) with 11 percent, and the Netherlands (Hoogovens) with 14 percent of the 7 million tons of total world supply of slab. Other countries supplying slab include West Germany (Thyssen), Sweden (SSAB), France (Usinor-Sacilor), and Australia (BHP). In the United States, the main suppliers of slab are Bethlehem (Sparrow Points), Geneva Works (formerly a USS plant), USS (Fairfield), and USS (Gary).

Of the twelve countries listed in Table 3.25, most countries, with the notable exceptions of Canada and Japan, were net exporters of semi-finished shapes in 1985. Canadian slab exports to the U.S. market fell, as the Canadians underwent capacity adjustment. U.S. imports from South Africa dropped because of the U.S. embargo on South African products.

Trends in semi-finished steel shapes trade differed among product groups (Table 3.26). Most of the growth in the international trade of semi-finished steel shapes during the 1977 to 1985 period was in blooms, billets, and slabs of carbon steel. International trade of carbon steel ingots oscillated without showing a clear trend. International trade of alloy and stainless steel

Table 3.22. U.S. Carbon Steel Slab Import by Country of Origin--1984 to July 1988

	1984	1985	1986	1987	1988
Slab Imports(thousand metric tons)	717.86	1,281.92	1,194.70	1,372.94	1,089.26
Percent Share Carbon Steel	99.80	98.70	99.50	99.60	99.70
Carbon Steel Slab (thousand metric tons)					
Canada	33.36	2.59	9.40	8.35	0.10
Mexico		0.92	47.34	35.97	27.64
Venezuela		37.47		36.12	24.46
Brazil	47.21	221.23	191.73	168.92	155.58
Sweden	105.33	247.65	227.11	253.77	184.81
Netherlands		3.01	6.62	29.85	10.01
Belgium	141.02	125.06	49.75	59.68	64.98
France	0.47	146.59	62.94	118.79	148.96
West Germany	371.01	365.27	209.94	213.65	153.36
Rumania		4.88			
South Korea	17.93	44.45	22.86	24.53	3.34
South Africa		49.82	56.13	20.48	
Japan			130.51	74.49	35.29
United Kingdom	0.28	16.26	174.66	269.26	194.80
Norway				20.83	
Australia					20.29
Other Countries	0.12	0.29		53.13	41.92
Total	716.74	1,265.51	1,189.01	1,367.00	1,086.37
Prices of Carbon Steel Slab (dollars per metric ton)					
Canada	235.41	245.37	217.76	218.71	278.21
Mexico		155.02	159.05	180.63	216.09
Venezuela		208.09		185.72	213.43
Brazil	231.26	198.49	192.90	185.67	214.00
Sweden	250.43	211.21	242.63	245.69	279.26
Netherlands		225.44	366.70	349.96	197.36
Belgium	197.35	198.49	193.73	210.32	248.95
France	238.63	219.46	219.56	245.10	262.44
West Germany	240.99	256.03	217.75	241.54	290.65
Rumania		309.21			
South Korea	198.10	204.09	222.37	196.06	183.71
South Africa		192.49	181.23	173.26	0.00
Japan			205.55	246.51	262.33
United Kingdom	363.91	219.18	204.56	200.16	208.74
Norway				238.64	
Australia					158.46
Other Countries	395.51	408.21		223.28	321.81

Source: U.S. Department of Commerce, *Imports by Country of Origin*;Various Years.

ingot, bloom, billets, and slabs first decreased and stayed at this lower level.

Table 3.23. Semi-Finished Steel Imports as Percent of Crude Steel Production in Selected Countries[a]-- 1977, 1979, 1980, 1983, and 1985 (percent)

Country	1977	1979	1980	1983	1985
Philippines	70	87	56	174	83
Turkey	13	3	4	9	26
Greece	12	41	30	52	41
Ecuador	n.a.	n.a.	716	139	567
Yugoslavia	11	11	9	7	9
Mexico	0	1	5	1	2
Spain	1	1	1	0	2
Bulgaria	1	0	0	0	6
Italy	2	3	3	2	4
Malaysia	1	58	57	161	43
Tunisia	29	3	0	9	22
Japan	0	1	0	0	0
Belgium	2	3	3	2	3
Nigeria	116	87	145	44	71
U.S.	0	0	0	1	3
Singapore	6	21	17	77	39
Austria	2	1	2	1	2
China	0	0	0	0	3

Note: [a] Includes all countries listed in Table 3.21.
Source: Iron and Steel Statistics Bureau, *World Steel Statistics Summary*, Various Years.

Another type of semi-finished steel trade arose (Table 3.27). Hot-rolled coil was long considered a finished product. It now often serves as an input to the cold-rolling mills and is classified as the semi-finished steel product hot-rolled coil for re-rolling.

In the 1977-85 period, international trade of hot-rolled coil for re-rolling grew from an average of 3 million tons per year in 1979-82 to an average of 6 million tons per year in 1983-85. From 1979 to 1985, the share of imports of hot-rolled coil for re-rolling from total imports of hot-rolled coil expanded from 19 percent to 34 percent. Imports of hot-rolled coil for re-rolling grew for each group of countries, with the big market countries accounting for the largest share of international trade in this product.

In the early 1970s, iron-rich countries planned to build large green-field steel plants with the objective of exporting semi-finished shapes (Table 3.28). The plants were to employ a conventional approach to steelmaking, involving coke plants, blast furnaces, basic oxygen furnaces, and continuous casting. They were to be owned and operated by multinational consortiums involving equity participation by the host government and the domestic steel industry. Suitable plant sites on deep water and near an available source of raw materials were considered essential.

Australia was often considered for semi-finished steel projects, mainly because it has abundant supplies of iron ore and coking coal. In the early

Table 3.24. Share of World Imports of Semi-Finished Steel for Selected Countries[a]--1977, 1979, 1980, 1983, and 1985 (percent)

Country	1977	1979	1980	1983	1985
Philippines	3	4	2	4	2
Turkey	3	1	1	4	10
Greece	1	3	3	5	3
Ecuador	1	1	1	0	1
Yugoslavia	5	4	4	3	3
Mexico	0	1	4	1	1
Spain	1	1	1	1	2
Bulgaria	0	0	0	0	1
Italy	5	8	10	4	7
Malaysia	0	1	1	7	2
Tunisia	1	0	0	0	0
Japan	1	5	5	2	3
Belgium	5	5	6	3	3
Nigeria	0	0	0	1	2
U.S.	5	4	1	11	16
Singapore	0	1	1	3	1
Austria	1	1	1	1	1
China	0	1	2	1	12

Note: [a] Includes all countries in Table 3.21.
Source: International Iron and Steel Institute, *Steel Statistics Yearbook,* Various Years and Iron and Steel Statistics Bureau, *World Steel Statistics Summary,* Various Years.

1970s, the Jumbo project in Western Australia, based at the Pilbara iron mines, attracted a number of big names in international steel, including Nippon Steel and Thyssen. This project, along with other semi-finished steel plants in Sweden and South Africa, was abandoned in the wake of the world recession and the steel crisis.

Only Brazil's Tubarao steel plant, which only produces semi-finished steel for export, was actually built. The plant is a joint-venture of Kawasaki (Japan), Finsider (Italy), and Siderbras (the Brazilian Steel Authority). Aside from Tubarao, Acominas was reported to be on stream, in 1988, to export semi-finished steel shapes, and plans to build a plant using the high grade iron ore from Carajas were announced. Large deposits of high grade iron ore, low labor costs, and efficient steelmaking facilities made Brazil a stronger competitor than other countries in the group.

International slab supply is not supposed to be a major constraint because new suppliers are expected to start production between 1989 and 1992. Major new projects are being developed or planned in Brazil, Venezuela, and Mexico, which may be targeted to the U.S. market.

Brazil is undertaking a pre-feasibility study of a new plant to export semi-finished steel. The Brazilian plant, Usimar, is expected to produce 3 million tons per year of slab or hot-rolled coil using conventional technology. The decision to produce slab or hot-rolled coil will depend on the studies. The

Table 3.25. Main Exporters of Semi-Finished Steel in 1987 (million metric tons)

Country	Total Exports of Semi-Finished Steel
Brazil	3.42
United Kingdom	1.65
Netherlands	1.35
West Germany	1.35
France	0.84
Belgium/Luxembourg	0.75
Spain	0.65
Australia	0.52
Sweden	0.39
Japan	0.25
South Korea	0.18
Canada	0.12

Source: Iron and Steel Statistics Bureau, "Selected Countries," 1988.

Usimar plant will be strongly competitive because of its strategic location at the port of Sao Luis in northern Brazil and its use of high-grade low cost iron ore from Carajas. The project attracted Russian interest which led to the signing of a letter of intent. Other interested parties include an Italian group (Italimpianti) that leads a consortium of French, British, and Czechoslovakian firms (*Metal Bulletin,* 1988). Brazilian economic problems have deferred implementation.

In 1985-86, Mexico and Venezuela reported plans to export semi-finished steel, and by 1986-88 these plans had been implemented. Both countries exported semi-finished steel mainly to the United States. The opening of Mexico's Sicartsa plant on the Pacific Coast in 1989 adds another major supplier. The Sicartsa plant is expected to produce 2 million tons of slab for export, mainly to the United States (Burnier, 1988).

Venezuela is now working on a pre-feasibility study of a new CVG plant which will have a capability of 1 million tons per year of slab and will be based in Guyana. This operation will employ direct reduction, EF furnaces, and continuous casting. The plant is estimated to cost $1 billion, and Kobe Steel (a Japanese steel producer) and Midrex (DRI equipment suppliers) are expected to take a maximum 49 percent share in the venture (*Metal Bulletin,* 1988).

Therefore, if Brazil's Tubarao plant does not lower its exports of slab and if all the projects under consideration are implemented, Brazil, Mexico, and Venezuela could be suppliers of 9 million tons per year of slab. This represents about 52 percent of the world total supply of semi-finished steel.

The increase through the 1980s of trade in semi-finished steel suggests that some situations exist in which such trade is profitable. Increasing semi-finished steel imports is attractive if the costs of producing the finished product are competitive. Chapter 6 further examines the extent to which such trade can be profitable.

Table 3.26. Imports Semi-Finished Product Mix by Regions[a] of the World--1977, 1979, 1980, 1983, and 1985

	1977	1979	1980	1983	1985
Imports of Slab- Carbon Steel (million metric tons)					
U.S.	0.02	0.09	0.00	0.16	0.59
E.E.C.	0.97	1.25	1.33	0.72	0.88
Japan	0.00	0.09	0.24	0.03	0.13
Big Market	0.99	1.43	1.57	0.92	1.59
Iron-Rich Industrial	0.01	0.03	0.04	0.02	0.03
Iron-Rich Developing	0.00	0.03	0.02	0.04	0.00
Total Iron-Rich	0.01	0.06	0.06	0.05	0.03
Oil-Rich Dev.Countries with Iron Ore[b]	0.15	0.14	0.12	0.11	0.02
Other Oil-Rich Dev. Countries	0.00	0.05	0.08	0.08	0.00
Total Oil-Rich Dev. Countries	0.15	0.19	0.20	0.18	0.02
Other Western	0.17	0.24	0.18	0.72	0.67
Total Western	1.31	1.91	2.01	1.87	2.31
C.P.E.	0.00	0.03	0.07	0.00	0.01
Unallocated	0.00	0.00	0.00	0.00	0.01
Total World	1.32	1.95	2.08	1.87	2.34
Imports of Bloom & Billet- Carbon Steel (million metric tons)					
U.S.	0.19	0.18	0.05	0.64	1.50
E.E.C.	1.25	1.52	1.39	0.94	1.41
Japan	0.00	0.15	0.18	0.10	0.25
Big Market	1.44	1.86	1.63	1.69	3.16
Iron-Rich Industrial	0.10	0.07	0.09	0.11	0.15
Iron-Rich Developing	0.00	0.03	0.03	0.01	0.09
Total Iron-Rich	0.10	0.11	0.12	0.12	0.24
Oil-Rich Dev.Countries with Iron Ore[b]	0.52	0.41	0.77	0.07	0.05
Other Oil-Rich Dev. Countries	0.22	0.32	0.55	0.76	0.65
Total Oil-Rich Dev. Countries	0.75	0.74	1.31	0.82	0.70
Other Western	1.05	1.16	1.30	1.91	3.11
Total Western	3.34	3.86	4.36	4.54	7.22
C.P.E.	0.00	0.09	0.01	0.13	1.95
Unallocated	0.47	0.10	0.13	0.56	0.15
Total World	3.81	4.04	4.50	5.23	9.31
Imports of Ingot- Carbon Steel (million metric tons)					
U.S.	0.13	0.04	0.06	0.14	0.08
E.E.C.	0.05	0.66	0.38	0.16	0.28
Japan	n.a.	0.18	0.01	0.00	0.00
Big Market[c]	0.17	0.89	0.45	0.30	0.37
Iron-Rich Industrial	0.01	0.12	0.07	0.00	0.01
Iron-Rich Developing	0.00	0.00	0.00	0.00	0.00
Total Iron-Rich	0.01	0.12	0.07	0.00	0.01
Oil-Rich Dev. with Iron Ore	0.00	0.00	0.01	0.00	0.10
Other Oil-Rich Dev. Countries	0.01	0.16	0.06	0.05	0.04
Total Oil-Rich Dev. Countries	0.01	0.16	0.08	0.05	0.14
Other Western	0.04	0.79	0.55	0.43	0.01
Total Western	0.23	1.96	1.15	0.78	0.53
C.P.E.	0.00	0.18	0.12	0.08	0.00
Unallocated	0.00	0.85	0.03	0.01	0.00
Total World	0.23	2.99	1.30	0.87	0.53

Table 3.26. (continued)

	1977	1979	1980	1983	1985
Other Semi-finished Steel (million metric tons)					
U.S.	0.05	0.05	0.01	0.02	0.05
E.E.C.	0.76	0.45	0.54	0.46	0.69
Japan	n.a.	0.08	0.00	0.00	0.00
Big Market	0.92	0.58	0.55	0.48	0.74
Iron-Rich Industrial	0.02	0.03	0.03	0.01	0.05
Iron-Rich Developing	0.00	0.01	0.01	0.01	0.01
Total Iron-Rich	0.02	0.04	0.04	0.02	0.05
Oil-Rich Dev. with Iron Ore	0.12	0.01	0.08	0.01	0.02
Other Oil-Rich Dev. Countries	0.14	0.00	0.00	0.00	0.00
Total Oil-Rich Dev. Countries	0.26	0.01	0.08	0.01	0.02
Other Western	0.73	0.23	0.29	0.16	0.59
Total Western	1.93	0.83	0.95	0.65	1.31
C.P.E.	0.29	0.05	0.04	0.00	0.01
Unallocated	0.56	0.00	0.01	0.00	0.00
Total World	2.78	0.88	1.00	0.66	1.32

Notes: [a] See text for definition of regions.
[b] Imports for Venezuela are not available for 1983 and 1984.
[c] Imports of ingots for Japan are not available for 1977.
Source: Iron and Steel Statistics Bureau, *World Steel Statistics Summary*, Various Years.

SUMMARY AND CONCLUSIONS

World steel output and consumption peaked at about 745 million tons in 1979, declined through 1982, and then rose to around 715 million tons in 1986. These developments are the net effect of declines in industrialized countries and increases in the CPE and developing noncommunist countries. Consumption and production in the industrialized countries peaked around 1973; growth tended to prevail from 1973 to 1979 in the rest of the world. Thus, 1979 marked the point at which the decline in the industrialized countries began to exceed rises elsewhere.

Since 1986 capacity in industrialized countries was at roughly the 1973 level, capacity utilization rates fell. A smaller decline in utilization rates prevailed for CPE countries; utilization rates rose for developing countries.

World trade also expanded over the 1973-86 period--from 114 million to 156 million tons according to export data and from 108 million to 156 million according to import data. The three main elements of the change were an 8 million-ton rise of U.S. net imports, a 7 million-ton fall in net imports by developing countries, and a 15 million-ton increase in net imports by the CPE countries (predominantly China and the U.S.S.R.). Brazil, South Korea, South Africa, and Rumania were the main sources of expanded exports. Another contribution came from iron-rich, oil-producing countries, mainly Venezuela and Mexico.

Table 3.27. Imports of Hot-Rolled Coil for Re-Rolling by Regions[a] of the World--1979, 1980, 1983, and 1985

	1979	1980	1983	1985
Imports of Hot-Rolled Coil for Re-Rolling				
U.S.	0.18	0.07	0.38	0.32
E.E.C.	1.75	1.75	1.95	1.71
Japan	0.00	0.00	0.03	0.10
Big Market[b]	1.93	1.81	2.37	2.13
Iron-Rich Indust. Countries	0.02	0.00	0.04	0.04
Iron-Rich Dev. Countries	0.03	0.07	0.02	0.06
Total Iron-Rich Countries	0.06	0.07	0.06	0.10
Oil-Rich Dev. Countries with Iron Ore	0.09	0.14	0.22	0.10
Other Oil-Rich Dev. Countries	0.04	0.05	0.59	0.50
Total Oil-Rich Dev. Countries	0.14	0.19	0.81	0.60
Other Western	0.49	0.72	2.05	1.73
Total Western	2.62	2.80	5.28	4.57
C.P.E.	0.11	0.06	0.55	0.64
Unallocated	0.00	0.00	0.00	0.00
Total World	2.72	2.86	5.83	5.20

Note: [a] See text for definition of regions.
Source: Iron and Steel Statistics Bureau, *World Steel Statistics Summary*, Various Years.

Examination of U.S. mill shipment data shows that all direct purchasers of steel reduced purchases from U.S. mills. A move to increased purchasing through service centers in the 1980s brought 1987 shipments to service centers to 18 million tons, compared to 18.5 million tons in 1973.

Mini-mills have arisen in the United States and have begun to account for the majority of U.S. industry shipments of non-flat-rolled products such as rods, bars, and structural shapes. Flat-rolled products still are produced predominantly by integrated mills; flat-rolled products, in turn, constitute the bulk of integrated-mill output.

The decline in steel production from 1979 levels has left 1986 world production of iron ore and pig iron and consumption of iron ore below 1979 levels. The much smaller sector producing direct reduced iron has expanded from 6.9 million tons in 1979 to 12.6 million tons in 1986.

The regional composition of iron ore, pig iron, and scrap supply also changed. The main trends in iron ore were increases in the CPE and iron ore-rich developing countries and declines in the U.S., the EEC, and the iron-rich industrialized countries. Brazil was the main source of iron-ore output rises. Since pig-iron trade changed little, pig-iron production trends tended to follow steel production.

Scrap generation comes predominantly from industrialized non-communist and CPE countries. Supplies declined in the industrialized noncommunist countries and remained steady in the CPE countries.

International trade of iron ore, scrap, semi-finished steel, and finished steel were higher in 1986 than in 1977. However, iron-ore exports peaked

Table 3.28. Semi-Finished Steel Plants Planned or Considered for 1980 (million metric tons of annualized capacity)

Country	Location	Initial Capacity
Brazil	Tubarao	3.0
Sweden	Lulea	4.0
South Africa	Saldanha Bay	3.0
Canada	Nova Scotia	3.0
Australia	West Australia	3.0

Source: J. Szekely, *The Future of the World's Steel Industry*, 1977, p. 18.

at 395 million tons in 1979 so that the 1986 level of 366 million tons were 7.5 percent below the 1979 level of 396 million tons. The main regional trends were a steady rise since 1973 of exports from iron-rich developing countries and a decline starting after 1979 that brought 1986 exports of iron-rich developed countries below their 1977 levels.

The markets with the largest percentage increases were scrap and semi-finished steel. Both markets had 1986 levels above 1979. The scrap market increased by 67 percent to 30 million tons in 1986. The U.S. is by far the largest net exporter of scrap and had by far the largest change in net exports.

The semi-finished market grew by 66 percent to 13.5 million tons when hot-rolled coil for re-rolling is excluded; the rise for trade including hot-rolled coil for re-rolling was 72 percent. Iron-ore trade increased by 3 percent and finished steel trade grew by 17 percent if hot-rolled coil for re-rolling is excluded and 19 percent if it is included. The main net sources are the EEC and the iron-rich developing countries, particularly Brazil. While the U.S. is the largest single net importer of semi-finished steel, U.S. net imports in 1986 were less than the total for the other-noncommunist-country group.

The chapters that follow emphasize the problems faced by the U.S. integrated producers and the role semi-finished steel products might have in alleviating these difficulties.

4

Adaption of the U.S. Steel Industry to Market Trends

The developments reviewed in Chapter 3 caused the financial performance of U.S. integrated steel producers to deteriorate. (See Appendix A for data on profitability and financing of new investments.) More critically, the experience suggested that difficulties would continue and, therefore, it was unprofitable to maintain then-existing levels of capacity. Extensive closings ensued.

This chapter deals with the aspects of this adjustment critical for analysis of the prospects for steel-slab imports. In particular, summary data are provided on capacity reduction and on the condition of the remaining capacity. This review is only part of the material gathered on the consequences of changing steel markets. Appendix B provides details on the capacity reduction summarized in this chapter.

The review deals successively with raw-material, coke-oven, blast-furnace, raw-steel, and finishing capacity. Particular attention is given to the Mountain and Pacific states. These regions never accounted for a large part of U.S. steel output, and the proportional loss of capacity in these western states exceeded that in the rest of the United States.

RAW MATERIAL PRODUCING CAPACITY REDUCTION

Capacity reduction arose in iron-ore and coal mining as well as at steel mills. U.S. iron-ore adjustments emphasized closing several high-cost mines. From 1973 to 1986, iron-ore production declined from 78 million to 38 million tons and imports of iron ore and concentrates declined from 44 million to 18 million tons. A large share of iron-ore imports originates in Canada. Although Canada's share of U.S. iron-ore imports increased from 50 percent to 54 percent, physical imports dropped 57 percent. In 1986, the reported Canadian iron-ore price was $9.41 per ton, more costly than other sources of imported iron ore. However, the mines are owned by integrated steel companies and probably operate because their actual variable costs delivered to U.S. mills are no greater than the delivered price of competitive ores.

Several steel companies such as LTV and Armco sold all their coal mines and signed long-term coking coal supply contracts. USS and Inland sold some of their coal mines (*World Steel Dynamics*, 1987). Presumably, this reflects the belief that coking coal supplies can be obtained economically on the open market.

STEEL PLANT CLOSINGS

Many different production facilities are housed in a steel plant. The shutdown process generally involves successively eliminating different parts of the plant. First, some equipment is allowed to wear down gradually and is scrapped when further maintenance is uneconomic. Eventually, the entire plant may be closed. Most capacity retirements in the 1980s involved eliminating only parts of plants. The data show that most of the closures were of plants producing bars and wire rods, the two products dominated by mini-mills (Barnett and Crandall, 1986).

Coke Ovens

Table 4.1 shows U. S. coke-oven capacities from 1973 to 1986. From 1977 to 1985, gross coking capacity declined from 62 million to 40 million tons, and coke production fell from 48 million to 30 million tons. Operating rates fell from 77 percent to 75 percent. Some observers of the steel industry have argued that aging capacity and the cost of its replacement would lead to increased coke imports. The data shown do not indicate any major increases in imports. From 1973 to 1978, coke imports increased from 1 million to 5 million tons; the latter level was less than 12 percent of production. Imports were well under one million tons in the 1980s. A low of 32,000 tons was reached in 1983; 1986 coke imports were about 300,000 tons (Table 4.1).

All major domestic producers reduced coking capacity (Tony, 1988). Pollution control requirements imposed on the coke ovens caused difficulties for integrated producers (McManus, 1984a). However, it is unclear how much of the capacity closing is due more to pollution control problems than to declining demand.

Given the age of the remaining coke batteries and the economics of replacement, coking capacity is expected to drop further. From 1989 to 1991, investments will be needed to replace 7 million tons of coke-oven capacity. These replacements may not occur, and further capacity decline may arise. Capital costs for rebuilding or maintaining present coke-oven capacities appear to be too high. The cost of rebuilding a coke oven is estimated at $125 per ton. A new coke oven costs around $250 per ton and takes two years to complete (U.S. International Trade Commission, 1988b; *World Steel Dynamics*, 1988).

Table 4.1. U.S. Coke Oven Capacity, Production, Operating Rate, and U.S. Coke Imports--1973 to 1986

	1973	1978	1979	1980	1983	1986
Coke Oven						
Capacity (mil.met.ton)	n.a.	60.84	61.81	58.02	41.17	n.a.
Production (mil.met.ton)	57.59	43.77	47.84	41.85	23.40	33.02
Operating Rate (percent)	n.a.	71.94	77.40	72.13	56.84	n.a.
Coke Imports (mil.met.ton)	0.98	5.19	3.61	0.60	0.03	0.30

Source: *World Steel Dynamics*, "Steel Strategist #14," 1987.

Coke ovens have a 25-year life expectancy. As of 1980, 47 percent of the batteries were over 20 years old and 25 percent were over 25 years old (*Iron Age*, 1984). According to a 1985 International Iron and Steel Institute survey, U.S. coke batteries are among the oldest in the industrialized countries with an average age of 19.3 years. The report concluded that the coke capacity of the noncommunist world would be sufficient through 1990. Beyond 1990, the report warned that the aging of many coke batteries could lessen coke supplies.

Domestic integrated producers could consider building new coke ovens, rebuilding old ones, arranging for third parties to construct coke plants, or adopting direct reduction iron-making methods. More coke might be imported. U.S. partnerships with foreign-owned companies abroad, for example, those in the EEC or Japan, are being considered (Tony, 1988). Alternatively, the high cost of domestic or foreign coke ovens might contribute to further reduction of U.S. steelmaking capacity and increased slab imports.

Blast Furnaces

From 1973 to 1986, U.S. blast-furnace capacity fell from 115 million to 75 million tons and pig-iron production declined from 94 million to 40 million tons. The blast-furnace operating rate dropped from 81 million to 53 percent. From 1977 to 1986, the number of blast furnaces withered from 182 to 85; those actually producing dropped from 115 to 42 (Table 4.2). Output of pig iron per blast furnace increased from 588,000 tons to 949,000 tons.

The number may continue to drop because of the unprofitability of investments in maintaining or replacing capacity. Extensive blast furnace relines and rebuilds are being deferred across the industry. Such deferrals can considerably raise the cost of rebuilding (McManus, 1984a; Soares, 1987).

Table 4.2. U.S. Blast Furnace Capacity, Production, Operating Rate, Coke Rate, and Number--1973 to 1986

		1973	1979	1980	1983	1986
Blast Furnace						
Capacity	(mil.met.ton)	115.34	113.05	105.06	91.69	73.20
Production	(mil.met.ton)	93.51	78.93	62.35	44.21	39.86
Operating Rate	(percent)	81.07	69.82	59.35	48.22	54.45
Number of Blast Furnaces		n.a.	173	149	109	85
Reported in Blast		159	117	83	49	42
Coke Rate	(kg/ton pig iron)	597	575	569	539	n.a.

Source: *World Steel Dynamics*, "Steel Strategist #14," 1987.

Crude Steel-Producing Capacity

Table 4.3 shows crude-steel capacity and production by process in the United States. The fullest data on capacity were available for the years from 1977 to 1986, a period in which U.S. steel-industry capacity fell from 145 million to 115 million tons.

The electric furnace (EF), for reasons discussed above, was the only process with increased capacity and production. From 1973 to 1986, the operating rate of U.S. electric furnaces declined from 92 percent to 65 percent, which is below the average operating rate of the noncommunist world (*World Steel Dynamics*, 1987).

Open-hearth (OH) furnace capacity and production dropped between 1973 to 1986. In 1986, OH furnaces accounted for 4 percent of the U.S. crude-steel production and 5 percent of crude-steel capacity. OH capacity was dismantled completely in the EEC and Japan by 1983 and 1978, respectively. In 1986, the share of OH production in the United States represented 22 percent of the noncommunist world OH production, while capacity accounted for 25 percent. From 1973 to 1986, the operating rate of OH declined from 98 percent to 51 percent.

Basic-oxygen furnace (BOF) production fell at a faster rate than capacity. As a result, the operating rate of the BOF declined from 99 percent to 64 percent. During the 1973-86 period, the BOF accounted for an increasing share of U.S. crude-steel production and capacity because of the dismantling of OH capacity.

Integrated-sector capacity increased until the early 1970s but subsequently declined. From 1970 to 1985, integrated-sector capacity dropped from 133 million to 101 million tons; conversely, mini-mill capacity expanded from 7 million to 20 million tons. The integrated sector produced 113 million tons in 1970 and 64 million tons in 1985; the mini-mill sector, 6 million and 16 million tons. The operating rates for the integrated sector declined from 85 percent to 63 percent, while those of the mini-mill fell from 93 percent to 81 percent (Barnett and Crandall, 1986).

During the 1977-86 period, integrated-producer capacity, including plants with only EF capacity and firms that stopped being fully integrated in

Table 4.3. U.S. Steelmaking and Continuous Casting Capacities, Production, and Operating Rates-- 1973 to 1986

	1973	1979	1980	1983	1986
Steelmaking Capacity (million metric tons)					
BOF	77.00	87.60	86.30	81.10	67.70
OH	37.00	18.90	15.50	12.40	5.90
EF	27.00	35.10	37.30	42.70	42.50
Production (million metric tons)					
BOF	75.90	75.90	62.00	47.50	43.30
OH	36.10	17.40	11.80	5.40	3.00
EF	24.80	30.40	27.60	22.70	27.40
Operating Rate (percent)					
BOF	98.60	86.60	71.80	58.60	64.00
OH	97.60	92.10	76.10	43.60	50.90
EF	91.90	86.60	74.00	53.20	64.50
Continuous Casting					
Capacity	14.45	26.37	27.93	38.03	56.94
Production	9.28	20.91	20.60	24.65	39.55
Operating Rate	64.20	79.30	73.80	64.80	69.50
Share of Steel Production	6.80	17.00	20.30	32.10	53.40

Source: *World Steel Dynamics*, "Steel Strategist #14," 1987.

1987, fell from 126 million to 89 million tons. Non-integrated producers increased their capacity from 19 million to 27 million tons (Table 4.4). As a result, the integrated producers' share of U.S. capacity fell from 87 percent to 77 percent, and the non-integrated producers' share rose from 13 percent to 23 percent (Table 4.4).

The EF capacity of integrated producers increased but not as quickly as in the mini-mills sector. The share of the U.S. EF capacity held by the mini-mills increased from 30 percent to 36 percent during the 1977-86 period. The integrated producers built up EF capacity to save declining plants, but even so, some plants were closed or sold (Barnett and Crandall, 1986).

The six largest producers--USS, LTV, Bethlehem, Inland, National, and Armco (all integrated producers)--reduced their share of the U.S. crude-steel capacity from 74 percent in 1977 to 61 percent in 1986 (Table 4.4). Partial and complete shutdown of plants and the spin-off of Weirton (1984) and Gulf States (1985) from National Steel and LTV Corporations, respectively, were major factors in the capacity reductions.

The share of the six largest integrated producers in U.S. steel shipments declined from 78 percent to 56 percent from 1960 to 1986 (Table 4.5). A market share around 60 percent could be attained if operating rates were above 80 percent.

Table 4.4. U.S. Crude Steel Capacity Ownership by Type of Firm--1977 to 1986

	Six Largest Producers	Integrated	Non-Integrated	Total
Capacity (million metric tons)				
1977	106.75	125.83	19.37	145.20
1978	103.88	123.96	19.46	143.42
1979	101.98	121.86	19.73	141.59
1980	99.72	117.71	21.28	138.99
1981	100.40	117.96	23.16	141.12
1982	97.32	114.92	24.71	139.63
1983	93.75	110.31	25.93	136.24
1984	82.61	96.52	26.50	123.02
1985	77.16	94.57	27.08	121.65
1986	71.13	89.09	26.79	115.88
Share (percent)				
1977	73.5	86.7	13.3	100.0
1978	72.4	86.4	13.6	100.0
1979	72.0	86.1	13.9	100.0
1980	71.7	84.7	15.3	100.0
1981	71.1	83.6	16.4	100.0
1982	69.7	82.3	17.7	100.0
1983	68.8	81.0	19.0	100.0
1984	67.2	78.5	21.5	100.0
1985	63.4	77.7	22.3	100.0
1986	61.4	76.9	23.1	100.0

Source: Compiled from *World Steel Dynamics*, "Core Report BB," 1987.

Continuous Casting

In 1977, continuous casting accounted for 12 percent of integrated-producer finishing capacity and 78 percent of mini-mill capacity. The corresponding 1986 ratios are 44 percent and 90 percent (*World Steel Dynamics*, 1987). From 1973 to 1986, U.S. continuous-casting capacity increased from 15 million to 57 million tons, and production of continuously cast steel increased from 9 million to 40 million tons. U.S. production of continuously cast steel rose from 7 percent to 54 percent. The noncommunist world continuously cast steel average increased from 12 percent to 67 percent (Table 4.3). The U.S. continuous-casting operating rate improved from 64 percent to 70 percent. Since the U.S. industry started to adopt the technology at a faster rate in 1984, operating rates should increase as managers of new plants gain experience about how best to employ the capacity.

Table 4.5. Share of the Six Largest Integrated Producers in the U.S. Steel Shipments--1950 to 1986

Year	Armco	Bethlehem	Inland	LTV	National	USS[a]	Total
1950	4.1	14.9	4.5	18.0	5.5	30.8	77.8
1960	6.7	15.3	6.8	16.6	7.1	25.0	77.5
1970	5.1	13.1	4.5	14.4	6.9	19.9	63.9
1982	6.1	13.2	6.7	8.3	5.6	16.2	56.1
1983	5.7	12.9	7.1	8.6	6.0	16.3	56.6
1984	5.6	12.1	6.8	11.2	6.1	16.0	57.8
1985	5.4	12.0	6.4	14.6	6.0	17.1	61.7
1986	5.4	12.1	7.0	13.1	6.4	12.1	56.0

Note: [a] USS was shut down after a labor disruption from August, 1986 to February, 1987.

Source: U.S. Congressional Budget Office, *The Effects of Import Quotas on the Steel Industry*, 1984, and Personal Interviews with Steel Producers and Trading Companies.

THE REGIONAL DISTRIBUTION OF THE CRUDE STEEL INTEGRATED CAPACITY

U.S. crude-steel producing capacity declined proportionally more in the West than in the East.[1] U.S. crude-steel capacity is concentrated in the East, especially in the Northeast. From 1977 to 1986, Eastern crude-steel capacity declined from 134 million to 112 million tons; in the West, from 11 million to 6 million tons. Thus, Eastern capacity dropped 17 percent; Western, 47 percent.

Eastern crude-steel production fell from 106 million tons in 1977 to 71 million tons in 1986, resulting in an operating rate decline from 79 percent to 63 percent. Western crude-steel production declined from 7 million to 3 million tons, and operating rates went from 66 percent to 55 percent. In the East, integrated producers' share of capacity declined from 89 percent to 79 percent, and in the West, from 68 percent to 31 percent. The more severe relative decline in the West is measured and further discussed in Appendix B.

CLOSING OF INTEGRATED MILL FINISHING CAPACITY AND MINI-MILL CAPACITY DEVELOPMENT

Most integrated-firm finishing capacity reduction occurred in the production of non-flat-rolled products. From 1977 to 1986, mini-mill

---

[1] The West here consists of the Pacific and Mountain states, and the East is the rest of the United States. See Appendices A and B for details.

producers increased their wire-rod and bar fabricating capacity while integrated producers reduced capacity (see Appendix B).

Table 4.6 shows U.S. mini-mills production, capacity, and utilization for steelmaking and for selected steel products of carbon and alloy steel from 1984 to 1988. Mini-mills capacity for producing reinforced bar, wire and related products, and structural shapes expanded since 1960. By 1986-87, more than 50 percent of the U.S. industry's capacity for making almost all of these products was in mini-mills. In 1987-88, the mini-mills' capacity share in these products (except wire and structural shapes) declined.

The most noticeable growth was in reinforced bar and wire products. Mini-mills capacity utilization rates were always above those for the total U.S. steel industry. Mini-mill shipments of these products increased in the period analyzed despite the decline of capacity and production share of the total U.S. industry (Table 4.7).

U.S. FLAT-ROLLED CAPACITY

Since the slabs transformed into flat-rolled products are all processed by reversing and hot-strip mills, the capacity of such mills measures the ability to finish steel into flat-rolled products. Table 4.8 presents total U.S. flat-rolled- product production capacity from 1977 to 1986. Flat-rolled-product production capacity declined from 88 million to 77 million tons during the years 1977 to 1986. Integrated producers decreased their share of flat-rolled capacity from 97 percent to 94 percent. The six largest producers reduced their share from 81 percent to 73 percent. The latter share exceeds that of the largest integrated producers in crude-steel production capacity. Other producers increased their share from 3 percent to 6 percent. (For further details, see Appendix B.)

In 1987-88, increased adoption of the electric furnace, wage concessions, and the rise in scrap prices narrowed the cost gap between integrated and mini-mill producers and improved the former's ability to compete in the non-flat-rolled product market.

The total or partial closures are an effort to eliminate marginal facilities. However, the survival potential of existing plants of integrated and non-integrated flat-rolled product producers is uncertain.

RESTRUCTURING AND THE COMPETITIVENESS OF EXISTING PLANTS

Despite these capacity reductions, U.S. capacity apparently remains less modern than that in other countries. One indicator is a rating by industry experts of the quality of carbon steel flat-rolled capacity and integrated hot-end facilities in 1988 (Table 4.9). This assessment is based on age, performance, and output quality. The ratings use the assignment of stars popularized in hotel and restaurant guides. The better rated facilities are assigned more stars. One- or two-star facilities also have limited life

Table 4.6. U.S. Mini-mills Production, Capacity and Utilization of Steelmaking for Selected Products of Carbon and Alloy Steel From 1984 to 1988[a]

	1984-85	1985-86	1986-87	1987-88
Steelmaking				
Electric Furnaces				
Percent of Total U.S. Capacity	43.2	44.1	50.1	50.0
Percent of Total U.S. Production	51.4	51.8	55.1	46.5
Capacity Utilization	72.6	80.0	80.7	85.9
Total Steelmaking				
Percent of Total U.S. Capacity	13.6	14.3	16.4	17.3
Percent of Total U.S. Production	15.7	16.8	19.9	17.1
Capacity Utilization	72.6	80.0	80.7	85.9
Continuous Casting				
Percent of Total U.S. Capacity	n.a.	31.9	28.0	29.0
Percent of Total U.S. Production	n.a.	32.3	30.6	27.5
Capacity Utilization	n.a.	77.8	82.8	82.9
Products				
Hot-finished Bars				
Percent of Total U.S. Capacity	39.7	41.1	50.0	45.6
Percent of Total U.S. Production	47.0	53.1	59.3	52.2
Capacity Utilization	62.0	77.5	75.0	81.2
Reinforced Bars				
Percent of Total U.S. Capacity	89.9	80.2	96.0	80.4
Percent of Total U.S. Production	89.0	87.5	98.0	82.1
Capacity Utilization	52.3	83.0	95.7	79.9
Wire Rod				
Percent of Total U.S. Capacity	39.4	45.8	53.8	49.4
Percent of Total U.S. Production	48.6	53.7	66.7	58.5
Capacity Utilization	84.4	84.4	96.2	96.1
Wire				
Percent of Total U.S. Capacity	38.9	28.8	57.9	59.2
Percent of Total U.S. Production	45.7	43.0	66.4	75.6
Capacity Utilization	63.2	74.6	61.7	79.2
Wire Products				
Percent of Total U.S. Capacity	64.9	81.0	81.3	77.8
Percent of Total U.S. Production	76.3	85.9	86.6	90.6
Capacity Utilization	61.3	64.6	77.5	72.8
Structural Shapes and Units				
Percent of Total U.S. Capacity	57.0	45.2	48.7	56.5
Percent of Total U.S. Production	67.2	52.5	62.6	62.7
Capacity Utilization	54.1	77.2	80.9	88.5

Note: [a] The period covered is from July 1st to June 30th of next year.

Source: U.S.International Trade Commission, *Annual Survey Concerning Competitive Conditions in the Steel Industry and Industry Efforts to Adjust and Modernize*, Report 2115, 1988.

Table 4.7. U.S. Mini-mills Producers' Shipments by Selected Products of Carbon and Alloy Steel from 1984 to 1988[a] (million metric tons)

Product	1984-85	1985-86	1986-87	1987-88
Semi-Finished	n.a.	0.37	0.68	0.67
Hot-Finished Bars	2.67	2.95	3.98	5.00
Reinforced Bars	2.79	3.46	3.13	2.18
Wire Rods	1.19	1.45	1.55	1.72
Wire	0.12	0.12	0.13	0.17
Wire Products	n.a.	n.a.	n.a.	n.a.
Structural Shapes and Units	2.12	2.12	2.29	2.90

Note: [a] The period covered is from July 1st to June 30th of next year.

Source: U.S. International Trade Commission, *Annual Survey Concerning Competitive Conditions in the Steel Industry and Industry Efforts to Adjust and Modernize*, Report 2115, 1988.

Table 4.8. Share of Largest Producers in U.S. Flat-Rolled Capacity--1977 to 1986

	Six Largest Producers	Integrated	Non-Integrated	Total
Capacity (million metric tons)				
1977	71.54	85.38	2.83	88.21
1978	69.85	83.69	2.83	86.52
1979	71.52	85.36	2.86	88.22
1980	67.85	81.83	2.86	84.69
1981	67.32	81.50	2.86	84.36
1982	67.23	81.41	2.86	84.27
1983	65.63	79.81	2.86	82.67
1984	62.55	74.28	2.96	77.24
1985	57.82	72.36	3.56	75.92
1986	56.08	72.38	4.56	76.94
Share (percent)				
1977	81.1	96.8	3.2	100.0
1978	80.7	96.7	3.3	100.0
1979	81.1	96.8	3.2	100.0
1980	80.1	96.6	3.4	100.0
1981	79.8	96.6	3.4	100.0
1982	79.8	96.6	3.4	100.0
1983	79.4	96.5	3.5	100.0
1984	81.0	96.2	3.8	100.0
1985	76.2	95.3	4.7	100.0
1986	72.9	94.1	5.9	100.0

Source: Compiled from *World Steel Dynamics*, "Steel Capacity Monitor #2," 1987.

expectancies. A facility generally can be upgraded one star with modest expenditures, but upgrading by two or more stars is likely to prove

Table 4.9. Aggregate U.S. Integrated Carbon Steel Industry Facility Rankings in 1988

	5 Stars[a]	4 Stars[b]	3 Stars[c]	2 Stars[d]	1 Star[e]	Total
Facility (million metric tons)						
Coke Ovens	1.50	4.99	8.12	7.61	2.86	25.07
Blast Furnaces	13.01	8.81	16.24	13.53	1.46	53.04
Steel Furnaces[f]	10.43	17.46	29.66	13.17	2.25	72.97
Slab Casters	7.26	25.80	4.67	4.81	0.00	42.54
Hot-Strip Mills	15.69	14.97	11.16	16.05	3.29	61.16
Plates Mills	0.00	1.63	2.77	2.04	0.27	6.71
Cold Reduction Mills	8.71	8.65	7.76	9.54	1.27	35.94
Hot-Dip Galvanizing	1.50	2.61	3.20	2.00	0.24	9.54
Electro-galvanizing	0.45	1.72	0.45	0.14	0.00	2.77
Tinplates	0.00	0.45	1.86	1.88	0.41	4.61
Shares (percent)						
Coke Ovens	6.0	19.9	32.4	30.3	11.4	100.0
Blast Furnaces	4.5	16.6	30.6	25.5	2.7	100.0
Steel Furnaces[f]	14.3	23.9	40.7	18.0	3.1	100.0
Slab Casters	17.1	60.7	11.0	11.3	0.0	100.0
Hot-Strip Mills	25.7	24.5	18.2	26.2	5.4	100.0
Plates Mills	0.0	24.3	41.2	30.4	4.1	100.0
Cold Reduction Mills	24.2	24.1	21.6	26.6	3.5	100.0
Hot-Dip Galvanizing	15.7	27.4	33.6	20.9	2.5	100.0
Electro-galvanizing	16.4	62.3	16.4	4.9	0.0	100.0
Tinplates	0.0	9.8	40.5	40.8	8.9	100.0

Note: [a] 5 stars, very high-rated facility 4 stars, high-rated facility.
[b] 4 stars, high-rated facility.
[c] 3 stars, medium-rating.
[d] 2 stars, low-rated facility.
[e] 1 star, very low-rated facility.
[f] Only integrated producers.
Ratings are based on age and general condition of facilities, performance, and acceptability of facility output.
Source: Barnett, *Steel Survival Strategies III*, 1988.

prohibitively expensive. A facility ranked as one star requires investment to upgrade or replace facilities. Some capital expenditures also are needed for two-star capacities. Most coke ovens and blast furnaces rate three stars or less. The proportion of hot-strip and tinplate mills with low ratings is even higher. The only facilities with at least 60 percent of capacity rated highly are the electro-galvanizing and slab casters. Most of these facilities were built recently, and none rate only one star. The large number of low-ranked facilities in the industry indicates that the modernization of the U.S. steel industry is not complete (Barnett, 1988).

SUMMARY AND CONCLUSIONS

The U.S. steel industry restructured its capacity in an attempt to reduce its production costs and accommodate the shifts in steel demand. U.S. integrated steel producers experienced a decline in their capacities. Mini-mills expanded their capacity and captured an increased share of the light non-flat-rolled product market. Therefore, capacity reduction is occurring primarily in integrated plants.

5

The Emergence of Imported Slabs in the U.S. Market and Its Implications for Integrated Steel Producers

Chapter 4 reviewed capacity reductions in the United States. Here the consequences of those reductions are examined. The characteristics of the U.S. flat-rolled product and semi-finished steel markets, trends, and the differences between the West and the East are stressed. The decline in the steelmaking capacity of the carbon-steel flat-rolled steel producers and its consequences are assessed; that appraisal is critical to this discussion. The material on import patterns provides background on the types of steel imported and the origins of these imports.

The United States is the largest flat-rolled-products-consuming country in the noncommunist world (U.S. International Trade Commission, 1988). U.S. steel industry restructuring lowered the capacity to produce semi-finished steel below the capacity to shape final products. Since some plants have more steel-making than finishing capacity, shipments have arisen from plants with extra steelmaking capacity to ones with less steelmaking than finishing capacity.

Given the overall disparity, such inter-plant shipments do not suffice to allow full utilization of finishing capacity. To provide additional inputs to finishing mills, the United States has thus become the world's largest semi-finished-steel-shape importing country. Steel demand in 1987 and 1988, for example, was large enough that surviving domestic capacity for producing semi-finished steel was inadequate and slab imports were increased.

THE FLAT-ROLLED AND SEMI-FINISHED STEEL MARKETS AND THEIR REGIONAL DIFFERENCES

This section examines market characteristics with special attention to the regional differences in the flat-rolled and semi-finished steel markets to identify the relevant steel grades and products that require the closest attention in subsequent sections.

The U.S. Flat-Rolled and Semi-Finished Steel Markets

Table 5.1 presents data by main product categories on the flat-rolled market. From 1973 to 1987, flat-rolled product consumption fell and imports grew, so that domestic output fell more than consumption. The import share of U.S. flat-rolled product consumption rose from 11 percent in 1973 to a peak of 22 percent in 1984. Subsequently, VER quotas and depreciation of the dollar helped reduce import penetration to its 1987 level of 17 percent.

Sheet and strip constitute the bulk of the flat-rolled product markets, i.e., about 91 percent of 1987 flat-rolled consumption; the rest is plate. Sheet and strip consumption fell by 18 percent; net shipments dropped by 25 percent. Import penetration rose from 11 percent to 17 percent. Plate consumption declined by 53 percent; net shipments, by 58 percent; imports, by 19 percent. Plate import penetration increased from 13 percent to 22 percent.

Exports of flat-rolled products fell by 68 percent with roughly equal percentage declines in sheet and strip and in plate exports. Net flat-rolled product imports grew by 49 percent. Net imports of plate peaked in 1978; sheet and strip net imports peaked in 1984. In 1987, net imports of plate were 4 percent below the 1973 level; net sheet and strip imports were 61 percent above the 1973 level.

Table 5.2 shows the mix among steel grades of domestic flat-rolled product supply and the share of gross shipments of steel producers transferred among steel-producing companies. The proportion of total flat-rolled product shipments transferred from company to company grew from 2.5 percent to 3.2 percent. The physical quantity dropped by 8 percent. Most of the steel transferred was sheet and strip products.

Carbon-grade steel accounts for at least 90 percent of all net shipment of flat-rolled products. The decline in net flat-rolled shipments was all in carbon steel; net stainless and alloy steel shipments increased from 1973 to 1987. The emphasis here, therefore, is on carbon steel flat-rolled shipments. Table 5.3 shows shipment, inter-plant transfer, import, and price data on all the major carbon steel flat-rolled product categories.

Most of the decline in flat-rolled carbon-steel net shipments occurred in plate and cold-rolled sheet and strip. A lesser drop prevailed for hot-rolled sheet and strip. Since sheet and strip coating results in the improved corrosion protection demanded by steel consumers, coated- and electrical sheet shipments rose, partially offsetting declines in shipments of other product types. The amount of carbon steel flat-rolled products transferred between plants of either the same or different companies increased.

From 1973 to 1987, sales revenue from flat-rolled carbon-steel products increased by 66 percent because the decline of shipments was less than the rise in nominal prices. However, sales revenues adjusted for inflation shrank by 27 percent because real composite flat-rolled prices declined from $324 to $287 per ton.

The largest real price drop occurred in coated and electrical sheets, followed by plates, hot-rolled sheet and strip, and cold-rolled sheet and strip. The amount of carbon-steel, flat-rolled products transferred between

Table 5.1. U.S. Shipments, International Trade, and Consumption of Flat-Rolled Products by Product Type--1979, 1980, 1983, 1986, and 1987

	1979	1980	1983	1986	1987
Import Penetration (percent)					
Plates	17.10	20.80	22.90	24.00	21.60
Sheet and Strip	12.90	11.40	16.30	17.70	16.50
Total Flat-Rolled Products	13.50	13.20	16.90	18.20	17.00
Plate (million metric tons)					
Net Shipments	8.19	7.33	3.46	3.23	3.67
Imports	1.65	1.87	1.00	1.00	0.99
Exports	0.19	0.19	0.09	0.06	0.08
Consumption	9.66	9.01	4.37	4.17	4.59
Net Exports	-1.46	-1.68	-0.91	-0.94	-0.92
Sheet & Strip (million metric tons)					
Net Shipments	45.18	35.65	35.46	36.72	38.51
Imports	6.53	4.40	6.83	7.78	7.50
Exports	0.91	1.38	0.42	0.46	0.52
Consumption	50.80	38.67	41.87	44.04	45.49
Net Exports	-5.61	-3.02	-6.41	-7.32	-6.98
Total Flat-Rolled Products (million metric tons)					
Net Shipments	53.38	42.98	38.92	39.96	42.18
Imports	8.18	6.27	7.83	8.79	8.49
Exports	1.10	1.57	0.51	0.52	0.59
Consumption	60.46	47.68	46.24	48.22	50.08
Net Exports	-7.08	-4.70	-7.32	-8.26	-7.90

Source: Compiled from American Iron and Steel Institute, *Annual Statistical Report,* Various Years.

plants of either the same or different companies increased from 3 million to 10 million tons, a rise of 5 percent to 21 percent in the manufacturers' shipments. Most of the flat-rolled products transferred between plants are sheet and strip.

Carbon-steel product imports grew until 1984 and then began to decline. From 1973 to 1987, a rise in net imports prevailed for all flat-rolled, carbon-steel products, except cold-rolled sheet.

Real and nominal prices of carbon-steel sheet and strip imports increased while real carbon-steel-plate prices dropped. The ratio of flat-rolled-product import prices to domestic prices followed a pattern that closely coincides with changes in U.S. dollar exchange rates with other major currencies (see Table 5.2 Chapter 6). Movements in the ratio between import and domestic prices closely interacted with exchange rate fluctuations. The price ratios rose from 1977 to 1980, fell from 1981 to 1983, and then began to move up again.

Exports of flat-rolled, carbon-steel products went down from 1973 to 1987, while nominal and real export prices increased. In 1987, exports rose as a result of the devaluation of the dollar. Nominal and real prices of

Table 5.2. U.S. Domestic Flat-Rolled Product Supply by Grade--1979, 1980, 1983, 1986, and 1987

	1979	1980	1983	1986	1987
Net Shipments by Steel Grade (million metric tons)					
Carbon Steel					
Plates	6.17	5.66	2.54	2.27	2.58
Sheet &Strip	42.23	33.18	32.81	33.79	36.03
Total Carbon	48.40	38.84	35.35	36.06	38.60
Alloy Steel					
Plates	1.89	1.55	0.83	0.86	0.95
Sheet &Strip	2.16	1.84	1.94	2.21	2.36
Total Alloy	4.06	3.39	2.77	3.07	3.31
Stainless Steel					
Plates	0.13	0.11	0.09	0.11	0.14
Sheet &Strip	0.79	0.64	0.71	0.73	0.85
Total Stainless	0.92	0.75	0.80	0.83	1.00
Shares of Steel Grades (percent)					
Carbon Steel	90.70	90.40	90.80	90.20	90.00
Alloy Steel	7.60	7.90	7.10	7.70	7.70
Stainless Steel	1.70	1.70	2.10	2.10	2.30
Transfer of Flat-Rolled Products Among Plants of Different Companies (million metric tons)					
Plates	0.02	0.02	0.00	0.01	0.01
Sheet and Strip	1.25	1.14	1.17	1.25	1.42
Total	1.31	1.15	1.20	1.24	1.42
Transfer of Flat-Rolled Products Among Plants of Different Companies (percent)					
Plates	0.30	0.30	0.10	0.30	0.20
Sheet and Strip	2.70	3.10	3.20	3.30	3.50
Total	2.40	2.60	3.00	3.00	3.20
Gross Shipments of Flat-Rolled Products (million metric tons)					
Plates	8.22	7.35	3.47	3.24	3.68
Sheet and Strip	46.45	36.78	36.65	37.96	40.65
Total	54.68	44.12	40.12	41.20	44.33

Source: Compiled from American Iron and Steel Institute, *Annual Statistical Report,* Various Years.

exported flat-rolled carbon-steel products, not only went up during 1973 to 1987, but have been above real and nominal import prices. This relationship may reflect increasing specialization on exports of more valuable grades of steel.

With dollar devaluation, flat-rolled exports rose, and imports from countries with currencies that had appreciated against the dollar fell. Table 5.4 shows the amount and share from 1973 to 1987 in U. S. flat-rolled product imports of the top eight supplying countries. In 1986-87, imports from Japan and the EEC countries fell, resulting in declining shares in total

Table 5.3. U.S. Shipments, Imports, and Prices of Carbon Flat-Rolled Carbon Steel by Type--1979, 1980, 1983, 1986, and 1987

	1979	1980	1983	1986	1987
Manufacturers' Shipments of Flat-Rolled Carbon Steel (million metric tons)					
Plate	6.50	5.91	2.40	2.33	2.91
Hot-Rolled Sheet & Strip	21.35	17.55	16.80	19.61	20.43
Cold-Rolled Sheet & Strip	16.71	12.86	13.09	12.57	12.68
Coated & Electrical Sheets	12.29	10.37	10.18	11.06	12.48
Total Flat-Rolled Products	56.86	46.68	42.46	45.58	48.50
Transfer of Carbon Steel Flat-Rolled Products Among Plants (million metric tons)[a]					
Plate	0.33	0.25	-0.14	0.06	0.33
Sheet and Strip	8.12	7.60	7.26	9.45	9.56
Total	8.46	7.84	7.11	9.52	9.90
Transfer of Carbon Steel Flat-Rolled Products Among Plants (percent)[a]					
Plate	5.10	4.20	-5.80	2.60	11.30
Sheet and Strip	16.10	18.60	18.10	21.90	21.00
Total	14.90	16.80	16.70	20.90	20.40
Nominal Price of Flat-Rolled Carbon Steel (dollars per metric ton)[b]					
Plate	431.83	473.14	432.61	398.11	423.63
Hot-Rolled Sheet & Strip	350.61	353.24	363.16	305.28	339.29
Cold-Rolled Sheet & Strip	447.25	456.58	510.10	488.94	477.14
Coated & Electrical Sheets	547.34	593.79	631.50	614.05	627.49
Total Flat Rolled Products	430.84	450.31	476.69	435.62	454.55
Real Price of Flat-Rolled Carbon Steel (1977 dollars per metric ton)[b] [c]					
Plate	355.85	341.92	277.40	258.03	267.37
Hot-Rolled Sheet & Strip	288.92	255.27	232.87	197.86	214.14
Cold-Rolled Sheet & Strip	368.55	329.95	327.08	316.90	301.15
Coated & Electrical Sheets	51.04	429.11	404.93	397.98	396.04
Total Flat-Rolled Products	355.04	325.42	305.66	282.34	286.88
Imports of Flat-Rolled Carbon Steel (million metric tons)					
Plate	1.61	1.84	1.24	1.36	1.42
Hot-Rolled Sheet & Strip	1.98	1.36	2.12	1.93	1.96
Cold-Rolled Sheet & Strip	2.15	1.33	2.17	2.38	2.01
Coated & Electrical Sheets	2.29	1.65	2.33	2.92	2.93
Total Flat Rolled Products	8.03	6.17	7.87	8.59	8.32
Imported Nominal Price of Flat-Rolled Carbon Steel (dollars per metric ton)[b]					
Plate	381.10	387.23	320.51	310.21	339.88
Hot-Rolled Sheet & Strip	359.40	359.15	63.95	334.36	364.91
Cold-Rolled Sheet & Strip	446.13	482.01	440.51	445.92	494.07
Coated & Electrical Sheets	542.33	595.45	564.68	562.12	599.09
Total Flat Rolled Products	439.14	469.21	430.39	438.86	474.42
Imported Real Price of Flat-Rolled Carbon Steel (1977 dollars per metric ton)[b] [c]					
Plate	314.05	298.37	204.53	201.05	214.51
Hot-Rolled Sheet & Strip	296.16	274.54	216.31	216.75	230.31
Cold-Rolled Sheet & Strip	367.64	348.32	282.46	289.01	311.83
Coated & Electrical Sheets	446.91	430.31	362.09	364.32	378.11
Total Flat-Rolled Products	361.88	339.08	275.98	284.44	299.42

Table 5.3 (continued)

	1979	1980	1983	1986	1987
Ratio of Imported and Domestic Prices[b]					
Plate	0.88	0.87	0.74	0.78	0.80
Hot-Rolled Sheet & Strip	1.03	1.08	0.93	1.10	1.08
Cold-Rolled Sheet & Strip	1.00	1.06	0.86	0.91	1.04
Coated & Electrical Sheets	0.99	1.00	0.89	0.92	0.95
Total Flat-Rolled Products	1.02	1.04	0.90	1.01	1.04
Exports of Flat-Rolled Carbon Steel (million metric tons)					
Plate	0.13	0.12	0.06	0.04	0.04
Hot-Rolled Sheet & Strip	0.07	0.10	0.03	0.03	0.05
Cold-Rolled Sheet & Strip	0.09	0.11	0.03	0.02	0.04
Coated & Electrical Sheets	0.12	0.18	0.07	0.06	0.10
Total Flat-Rolled Products	0.41	0.50	0.19	0.15	0.23
Exported Nominal Price of Flat-Rolled Carbon Steel (dollars per metric ton)[b]					
Plate	403.27	441.36	335.59	529.56	657.82
Hot-Rolled Sheet & Strip	389.09	391.72	588.89	522.42	481.84
Cold-Rolled Sheet & Strip	548.73	571.33	869.11	1034.32	967.07
Coated & Electrical Sheets	620.49	620.45	785.37	930.39	808.91
Total Flat-Rolled Products	497.00	522.51	631.70	768.79	732.77
Exported Real Price of Flat-Rolled Carbon Steel (1977 dollars per metric ton)[b][c]					
Plate	332.31	318.95	215.19	343.22	415.17
Hot-Rolled Sheet & Strip	320.63	283.08	377.61	338.60	304.11
Cold-Rolled Sheet & Strip	452.18	412.87	557.29	670.37	610.36
Coated & Electrical Sheets	511.32	448.37	503.60	603.00	510.53
Total Flat Rolled Products	409.55	377.59	405.06	498.27	462.48
Ratio of Exported and Domestic Prices[b]					
Plate	0.93	0.93	0.78	1.33	1.55
Hot-Rolled Sheet & Strip	1.11	1.11	1.62	1.71	1.42
Cold-Rolled Sheet & Strip	1.23	1.25	1.70	2.12	2.03
Coated & Electrical Sheets	1.13	1.04	1.24	1.52	1.29
Total Flat-Rolled Products	1.15	1.16	1.33	1.76	1.61

Notes: [a] Includes quantities of steel consumed in steel-producing plants in the manufacture of fabricated and as maintenance, repair, and operating supplies.

[b] U.S. import prices prior to 1976 were customs values, after 1976 tariffs, insurance, and freight are included. Therefore, prices from 1973 to 1976 are not comparable with prices from 1977 to 1987.

[c] Prices converted to constant dollars using the GNP deflator rate from 1977 to the year in question.

Source: Compiled from *Current Industrial Reports, Steel Mill Products*, MA33B, U.S. Department of Commerce, Various Years.

flat-rolled imports. The share of flat-rolled imports from Canada, Brazil, and South Korea increased, since this devaluation of the dollar mostly affected the EEC and Japan.

Rises in total flat-rolled import penetration from 1973 to 1987 were, however, less than the move from 12 percent to 21 percent of the ratio of

Table 5.4. Role of Leading Suppliers[a] in U.S. Flat-Rolled Product Imports--1979, 1980, 1983, 1986, and 1987

	1979	1980	1983	1986	1987
Flat-Rolled Imports by Country of Origin (million metric tons)					
Japan	2.86	2.25	2.18	2.28	2.22
Canada	0.69	0.65	0.70	1.03	1.42
West Germany	1.37	0.76	0.84	1.15	0.95
South Korea	0.34	0.33	0.61	0.68	0.61
France	0.87	0.59	0.52	0.58	0.56
Brazil	0.22	0.30	0.76	0.36	0.38
Netherlands	0.44	0.30	0.27	0.36	0.33
Belgium	0.33	0.37	0.24	0.31	0.23
Share of Flat-Rolled Imports by Country of Origin (percent)					
Japan	35.00	35.90	27.80	25.90	26.10
Canada	8.50	10.40	8.90	11.80	16.70
West Germany	16.70	12.10	10.80	13.10	11.10
South Korea	4.10	5.20	7.80	7.80	7.20
France	10.70	9.40	6.70	6.60	6.50
Brazil	2.70	4.80	9.70	4.10	4.50
Netherlands	5.30	4.70	3.40	4.10	3.90
Belgium	4.00	6.00	3.10	3.50	2.70

Note: [a] Includes eight largest suppliers in 1987.
Source: Compiled from American Iron and Steel Institute, *Annual Statistical Report,* Various Years.

imports to total steel mill product consumption. The main source of this rising overall import share was an increase in the share of semi-finished steel imports in consumption from 5 percent to 60 percent. Table 5.5 shows the characteristics of the semi-finished steel market, which includes ingot, bloom, billet, and slab. Domestic net shipments of semi-finished steel dropped 57 percent; consumption remained at 3 million tons. As a result, imports of semi-finished steel expanded from almost nothing to 2 million tons, and the United States became a net importer of semi-finished steel. The ratio of imports to manufacturers' shipments rose from 1 percent to 33 percent; manufacturers' shipments declined by 64 percent (Tables 5.5 and 5.6). The ratio exceeds that for total steel-mill-product imports.

Transfer of semi-finished products among plants of different companies varied from 1973 to 1987. Basically, it declined from 1973 to 1982 but recovered and began to rise, especially in 1987, reaching 52 percent of gross shipments. Net semi-finished product shipments are less heavily concentrated in carbon-steel grades than are flat-rolled-product shipments. Semi-finished carbon steel accounts for slightly more than 60 percent of net shipments.

The most significant decline in semi-finished steel products was in shipments among plants of the same company. The manufacturers' shipments of semi-finished carbon steel fell by 70 percent, accounting for

Table 5.5. U.S. Shipments, International Trade, and Consumption of Semi-Finished Steel--1979, 1980, 1983, 1986, and 1987 [a]

	1979	1980	1983	1986	1987
Import Penetration (percent)[b]					
Semi-Finished Steel		13.20	8.20	48.10	59.40	60.00
Total Steel Mill Products		15.40	16.30	19.90	21.90	20.50
Ratio of Imported and U.S. Gross Shipments					
Semi-finished Steel			2.60	1.50	15.60	41.50	33.00
Semi-Finished Steel Market (million metric tons)					
Net Shipments	2.36	2.39	0.91	1.35	1.44
Imports	0.31	0.14	0.75	1.89	2.07
Exports	0.32	0.83	0.09	0.05	0.07
Consumption	2.35	1.70	1.56	3.18	3.45
Net Exports	0.01	0.69	-0.65	-1.84	-2.00
Net Shipments of Semi-Finished Steel by Grade (million metric tons)					
Carbon Steel	1.37	1.67	0.48	0.89	0.88
Alloy Steel	0.93	0.67	0.39	0.41	0.50
Stainless Steel	0.06	0.05	0.03	0.04	0.07
Total	2.36	2.39	0.91	1.35	1.45
Share of Steel Grades (percent)					
Carbon Steel	58.10	69.90	52.90	66.10	60.60
Alloy Steel	39.30	27.90	43.60	30.50	34.50
Stainless Steel	2.60	2.20	3.50	3.30	4.80
Transfer of Semi-Finished Steel Among Plants of Different Companies (million metric tons)					
All Grades	0.22	0.13	0.05	0.32	1.59
Transfer of Semi-Finished Steel Among Plants of Different Companies (percent)					
All Grades	8.50	5.00	4.80	19.20	52.40
Gross Shipments of Semi-Finished Steel (million metric tons)					
All Grades	2.58	2.51	0.95	1.67	3.03

Notes: [a] Includes only ingot, bloom, billet, and slab.
[b] The ratio of import to consumption quantities.
Source: Compiled from American Iron and Steel Institute, *Annual Statistical Report,* Various Years.

almost all the fall of semi-finished steel shipments. Semi-finished alloy steel dropped by 24 percent, while the stainless-steel manufacturers' shipments remained almost constant throughout the period covered. The largest reduction in the domestic supply of carbon and alloy steel took place in 1981-82, when total steel supplies went from 9 million to 4 million tons.

Table 5.6 shows total semi-finished steel trade among steel producers and the quantity and average annual prices for manufacturers' shipments, imports, and exports. At least 70 percent of total manufacturers' shipments are traded among steel plants of either the same or different companies. This share varies for different grades of steel. The shares for carbon and

Table 5.6. U.S. Shipments, Imports, and Prices of Semi-Finished Steel by Grade--1979, 1980, 1983, 1986, and 1987 [a]

	1979	1980	1983	1986	1987
Manufacturer's Shipments of Semi-Finished Steel (million metric tons)					
Carbon Steel	10.03	7.38	3.75	3.11	4.33
Alloy Steel	1.76	1.60	0.68	0.83	1.18
Stainless Steel	0.35	0.22	0.39	0.24	0.28
Total	12.14	9.20	4.82	4.17	5.78
Total Transfer of Semi-Finished Steel Among Plants (million metric tons)[b]					
Carbon Steel	8.66	5.71	3.27	2.22	3.45
Alloy Steel	0.83	0.93	0.29	0.42	0.68
Stainless Steel	0.29	0.17	0.36	0.20	0.21
Total	9.78	6.81	3.91	2.82	4.33
Total Transfer of Semi-Finished Steel Among Plants (percent)[b]					
Carbon Steel	86.30	77.40	87.20	71.40	79.70
Alloy Steel	47.20	58.10	42.60	50.60	57.60
Stainless Steel	82.90	77.30	92.30	83.30	75.00
Total	80.60	74.00	81.10	67.60	74.90
Domestic Nominal Price of Semi-Finished Steel (dollars per metric ton)[c]					
Carbon Steel	308.33	336.97	311.96	237.07	294.21
Alloy Steel	626.99	705.43	727.31	658.51	622.13
Stainless Steel	1440.38	1574.23	1307.08	1336.25	1256.67
Domestic Real Price (1977 dollars) of Semi-Finished Steel (dollars per metric ton)[c][d]					
Carbon Steel	254.08	243.51	200.04	153.65	185.69
Alloy Steel	516.67	509.78	466.37	426.79	392.65
Stainless Steel	1186.95	1137.63	838.13	866.05	793.14
Imports of Semi-Finished Steel (million metric tons)					
Carbon Steel	0.24	0.11	0.71	1.63	1.76
Alloy Steel	0.07	0.03	0.03	0.08	0.09
Stainless Steel	0.00	0.00	0.00	0.02	0.05
Total	0.31	0.14	0.75	1.73	1.91
Imported Nominal Prices of of Semi-Finished Steel (dollars per metric ton)[c]					
Carbon Steel	260.97	285.80	247.68	225.46	245.95
Alloy Steel	619.11	752.81	602.15	558.15	536.04
Stainless Steel	1277.56	1843.46	1617.31	1193.71	926.61
Imported Real Prices (1977 dollars) of Semi-Finished Steel (dollars per metric ton)[c][d]					
Carbon Steel	215.05	206.54	158.82	146.13	155.23
Alloy Steel	510.18	544.02	386.11	361.75	338.32
Stainless Steel	1052.78	1332.19	1037.05	773.67	584.82
Ratio Imported and Domestic Prices[c]					
Carbon Steel	0.85	0.85	0.79	0.95	0.84
Alloy Steel	0.99	1.07	0.83	0.85	0.86
Stainless Steel	0.89	1.17	1.24	0.89	0.74
Exports of Semi-Finished Steel (million metric tons)					
Carbon Steel	0.24	0.54	0.02	0.04	0.03
Alloy Steel	0.08	0.28	0.07	0.01	0.04
Stainless Steel	0.00	0.00	0.00	0.00	0.00
Total	0.32	0.83	0.09	0.05	0.07

Table 5.6 (continued)

	1979	1980	1983	1986	1987
Exported Nominal Prices of of Semi-Finished Steel (dollars per metric ton)c					
Carbon Steel	244.42	266.65	291.51	225.75	390.31
Alloy Steel	356.10	343.23	268.43	646.53	507.42
Stainless Steel	3395.40	2009.90	1969.85	2253.66	2466.42
Exported Real Prices (1977 dollars) of Semi-Finished Steel (dollars per metric ton)c d					
Carbon Steel	201.42	192.70	186.92	146.31	246.34
Alloy Steel	293.45	248.04	172.12	419.03	320.25
Stainless Steel	2797.99	1452.47	1263.11	1460.65	1556.66
Ratio of Exported and Domestic Pricesc					
Carbon Steel	0.79	0.79	0.93	0.95	1.33
Alloy Steel	0.57	0.49	0.37	0.98	0.82
Stainless Steel	2.36	1.28	1.51	1.69	1.96

Notes: a Includes only ingot, bloom, billet and slab.

b Includes quantities of steel consumed in steel-producing plants in the manufacture of fabricated steel and as maintenance, repair, and operating supplies.

c U.S. import prices prior to 1976 were customs values; after 1976 tariffs, insurance, and freight are included. Therefore, prices from 1973 to 1976 are not comparable with prices from 1977 to 1987

d Prices converted to constant dollars using the GNP deflator rate from 1977 to the year in question.

Source: Compiled from *Current Industrial Reports, Steel Mill Products*, MA33B, U.S. Department of Commerce, Various Years.

stainless-steel shipments average almost 80 percent; the share of alloy steel went from 34 percent to 58 percent from 1973 to 1987. Nominal prices of domestic supplies increased for all steel grades. However, real prices for alloy and stainless decreased. The real price of carbon steel fell from 1973 to 1986, but in 1987 was above the 1973 price. A tight market for this grade of steel occurred, and imports were limited by the VER.

As the ability of domestic steel producers to supply the U.S. market declined, imports of semi-finished steel rose. Growth was mainly in carbon steel grade imports, which increased consistently from 0.1 million to 1.8 million tons. Imports of the other grades expanded, but the total amount of alloy and stainless imports was not as significant as that for carbon steel.

Nominal prices for all grades of imported steel rose from 1977 to 1987, while a real price rise was only observed for alloy steel. Real prices for carbon and stainless dropped during the period examined. However, the ratio between imported and domestic prices declined for all grades. The most critical fall was in carbon steel. Therefore, the nominal prices of imported supplies increased less than that of domestic supplies, stimulating higher semi-finished carbon-steel imports.

Exports of semi-finished shapes fell more than did manufacturers' shipments. From 1977 to 1987, exports of semi-finished steel dropped by

Table 5.7. U.S. Semi-Finished Steel Imports by Country of Origin--1979, 1980, 1983, 1986, and 1987

	1979	1980	1983	1986	1987
Semi-Finished Steel Imports by Country of Origin (million metric tons)					
Brazil	0.05	0.01	0.04	0.37	0.43
Sweden	0.07	0.01	0.05	0.28	0.31
United Kingdom	0.07	0.02	0.01	0.20	0.30
West Germany	0.00	0.00	0.12	0.24	0.27
France	0.00	0.00	0.02	0.11	0.14
Canada	0.05	0.09	0.40	0.19	0.12
Japan	0.00	0.00	0.00	0.14	0.08
Netherlands	0.00	0.00	0.04	0.07	0.08
Belgium	0.00	0.00	0.00	0.07	0.06
Mexico	0.00	0.00	0.00	0.06	0.06
Venezuela	0.00	0.00	0.00	0.00	0.04
South Korea	0.06	0.00	0.00	0.03	0.02
Italy	0.01	0.00	0.00	0.01	0.01
South Africa	0.00	0.00	0.04	0.06	0.02
Share of Semi-Finished Steel Imports by Country of Origin (percent)					
Brazil	17.20	8.20	5.00	19.50	20.80
Sweden	21.00	7.50	7.10	14.80	15.00
United Kingdom	21.60	12.40	1.30	10.40	14.60
West Germany	0.90	1.80	16.00	12.70	12.90
France	0.40	0.20	2.90	5.90	6.90
Canada	15.10	66.10	53.30	10.20	5.90
Japan	1.10	2.60	0.20	7.60	4.10
Netherlands	0.00	0.00	5.00	3.50	3.90
Belgium	0.20	0.20	0.10	3.60	2.90
Mexico	0.00	0.00	0.10	3.20	2.70
Venezuela	0.00	0.00	0.00	0.00	1.70
South Korea	20.50	0.00	0.00	1.50	1.20
Italy	1.80	0.40	0.10	0.60	0.60
South Africa	0.20	0.00	4.80	3.00	1.00

Source: Compiled from American Iron and Steel Institute, *Annual Statistical Report*, Various Years.

86 percent. Mainly, the decline in exports of semi-finished shapes resulted from domestic prices being higher than export prices (Table 5.7).

Table 5.7 lists the import share for semi-finished steel from 1973 to 1987, by country of origin. In 1973, imports were almost nonexistent. Starting in 1984, imports of semi-finished steel became larger and more diversified. Data on all suppliers are available only from 1978 on. In 1978, the top five suppliers accounted for 95 percent of the imported of semi-finished steel. In 1987, the top five suppliers provided 70 percent of these imports; the top eight, 84 percent. Of the five leading suppliers of 1978, only Brazil increased its share of U.S. imports. The top eight countries' share of imported semi-finished steel exceeds that for flat-rolled products (79 percent). As new suppliers (see Chapter 3) develop, the geographic

84   Implications of Imported Slabs in the U.S. Market

concentration of imported supplies of semi-finished steel will decrease further.

Developing countries increased their share of the U.S. semi-finished steel import market. Brazil became the largest source for imported semi-finished steel. Mexico and Venezuela enlarged their shares during 1987-88 and are expected to account for an even larger share in the future.

Sweden, Canada, and South Africa had decreased shares. The Swedish share declined as its shipments grew until 1984 and remained at that amount while total imports continued to expand. Canadian shipments fell from 1981 to 1987, as a result of its capacity restructuring. South African imports are restricted by the Anti-Apartheid Act of 1986.

EEC shipments to the United States increased from 1973 to 1987. From 1985 to 1987, the imports from the United Kingdom and the Netherlands rose, while those from other EEC countries, especially West Germany, fell. This trend is associated with the devaluation of the dollar. Japan sporadically exported semi-finished steel to the United States; shipments only reached significant levels in 1986-87. Therefore, Japan cannot as yet be considered a regular supplier.

*Regional Differences in the Flat-Rolled
Carbon Steel and Semi-Finished Steel Markets*

Flat-rolled carbon steel products and semi-finished steel imports increased in both the East and the West (Table 5.8). From 1973 to 1987, plate imports declined and sheet and strip imports increased. Data from Table 5.8 are not sufficient because the table does not show the vulnerability of each region to flat-rolled imports and their dependence on semi-finished steel imports.

To determine the differences between the East and the West, indirect measures were used to estimate import penetration because regional domestic-consumption data were unavailable (Table 5.9). The available data only permitted showing this difference for flat-rolled products from 1977 to 1986.

From 1977 to 1986, the proxy used to estimate flat-rolled import penetration (the ratio of flat-rolled imports to U.S. flat-rolled capacity) was larger in the West than in the East. The proxy increased from 9 percent to 10 percent in the East, while in the West it rose from 39 percent to 45 percent. This difference was consistent among individual flat-rolled products categories. Therefore, import penetration was higher in the West than either the East or the total U.S. market. The greatest import penetration occurred in 1984 when the Kaiser plant closed. In 1985, the proxy declined because the plant resumed operations as CSI, relying completely on slab purchases.

From 1977 to 1987, the proxy used to measure import penetration of semi-finished steel (the ratio of imports of semi-finished shapes to U.S. crude-steel production) for the U.S. market rose from almost nothing to 3 percent. Again the West experienced a larger import penetration than did the East. In 1985, when the CSI plant became dependent upon slab imports,

Implications of Imported Slabs in the U.S. Market

Table 5.8. U.S. Imports of Selected Steel Products by Region--1979, 1980, 1983, 1986, and 1987 (million metric tons)

	1979	1980	1983	1986	1987
Ingot, Bloom, Billet, and Slab					
East	0.31	0.14	0.63	1.48	1.63
West	0.00	0.00	0.11	0.41	0.44
Plate					
East	1.46	1.57	0.82	0.85	0.87
West	0.19	0.30	0.19	0.16	0.12
Sheet & Strip					
East	5.05	2.88	5.06	6.19	5.86
West	1.47	1.52	1.77	1.59	1.63
Flat-Rolled Products					
East	6.51	4.45	5.87	7.04	6.74
West	1.67	1.82	1.95	1.75	1.75
Total Steel Mill Products					
East	12.44	10.36	11.60	14.59	14.34
West	3.44	3.69	3.89	4.17	4.17

Source: Compiled from American Iron and Steel Institute, *Annual Statistical Report*, Various Years.

import penetration in the West jumped from below the U.S. average to 13 percent. Western import penetration was 15 percent, compared to the U.S. average of 3 percent in 1987.

Therefore, the West is, not only more vulnerable to flat-rolled import penetration than the East, but more dependent on semi-finished steel imports. The data on regional capacity differences in Chapter 4 and Appendix B imply increased Western problems. The West has no continuous casting capacity, and its crude-steel-producing plant relies on OH furnaces.

THE DEFICIENCY OF THE STEELMAKING CAPACITY OF U.S. FLAT-ROLLED PRODUCERS AND ITS IMPACT ON MARKET STRUCTURE

This section covers the reliance on carbon-steel slabs imports and their share in total semi-finished steel imports. Data on this topic are very limited. The section also shows the effects of the reduction in steelmaking capacity on the U.S. flat-rolled producers and its impact on market structure.

The deficiency in the steelmaking capacity of the U.S. integrated producers emerged in 1982. Crude-steel-producing capacity declined more rapidly than finishing capacity, and the demand for semi-finished steel began to rise. However, this drop of steelmaking capacity received little attention due to the low operating rate of the U.S. steel industry. As the

Table 5.9. Estimated Ratio of Import to Domestic Production of Flat-Rolled and Semi-Finished Products by Region and Product Type--1979, 1980, 1983, 1986, and 1987

	1979	1980	1983	1986	1987
East					
Semi-Finished[a]	0.27	0.14	0.87	2.09	2.09
Plates[b]	13.05	14.48	7.69	10.37	n.a.
Sheet & Strip[c]	6.28	4.17	7.53	9.55	n.a.
Flat-Rolled Products[d]	7.79	5.57	7.56	9.64	n.a.
Total Steel Products[e]	10.76	10.93	15.92	20.62	18.38
West					
Semi-Finished[a]	0.06	0.06	2.90	12.58	15.60
Plates[b]	16.66	25.62	15.86	37.47	n.a.
Sheet & Strip[c]	42.58	42.25	46.49	45.36	n.a.
Flat-Rolled Products[d]	36.03	38.17	39.28	44.52	n.a.
Total Steel Products[e]	42.95	54.89	99.90	128.46	148.72
U.S. Market					
Semi-Finished[a]	0.25	0.14	0.97	2.55	2.56
Plates[b]	13.40	15.57	8.50	11.70	n.a.
Sheet & Strip[c]	8.60	6.06	9.62	11.38	n.a.
Flat-Rolled Products[d]	9.27	7.41	9.46	11.42	n.a.
Total Steel Products[e]	12.85	13.84	20.18	25.36	22.90

Notes: [a] Ratio of ingot, bloom, billet, and slab import and domestic crude steel production.
[b] Ratio of plate import and reversing mill capacity.
[c] Ratio of sheet and strip import and hot-strip mill capacity.
[d] Ratio of plate plus sheet and strip import and total flat-rolled capacity.
[e] Ratio of total steel mill products and crude steel production.

Source: Compiled from *World Steel Dynamics*, "Capacity Monitor #2" and "Core Report BB," 1987 and American Iron and Steel Institute, *Annual Statistical Report,* Various Years.

operating rate started to recover from a 1982 low, the inadequacy of steelmaking capacity became more noticeable, especially in 1987-88 when operating rates were above or equal to 80 percent.

To measure disparities between steelmaking and finishing capacities, I used capacity balance equations (see Appendix C for the equations used). These equations were checked by a steel industry analyst with whom I consulted. The approach takes into account the steelmaking and finishing capacities, slab continuous-casting capacity, process transformation yields and crude-steel distribution among flat-rolled and non-flat-rolled capacities. The calculations were implemented using the data from *World Steel Dynamics*. The details of the calculations cannot be disclosed due to

Implications of Imported Slabs in the U.S. Market

information confidentiality; however, the results produced by its use are reported here.

The flat-rolled market is the focus of the computations. The steelmaking and slab continuous-casting capacities of different plants and the reversing mill and hot-strip mill capacities are the critical data in determining what is needed to attain the desired output. The product mix in 1986 and the distribution of flat-rolled and non-flat-rolled capacity in 1987 were used to allocate crude-steel production to flat-rolled product production. This is necessary since some capacity is used to serve the production of non-flat-rolled products.

The calculations were made allowing a capacity utilization rate for steelmaking, finishing mills, and continuous-casting facilities of no more than 90 percent. Steel producers consider 90 percent utilization the maximum regularly attainable. The capacities of all producers having steelmaking facilities until 1986 were included.

Tables 5.10 and 5.11 show the difference between producing capacity and desired flat-rolled product output at different flat-rolled product mill capacity utilization rates for the years 1977 to 1986. Table 5.10 considers the product mix of steel producers and the total steelmaking capacity of all their plants because ingots can be produced at non-flat-rolled plants and shipped to plants with insufficient slab capacity. Table 5.11 only covers plants producing flat-rolled products and their distribution of flat-rolled and non-flat-rolled products producing capacity.

The design of these tables is such that the difficulties in meeting production goals necessarily are greater when a higher flat-rolled product mill capacity utilization rate is assumed and when only steelmaking capacity at plants with flat-rolled producing finishing capacity is considered. The only results that are determined by the data are the time pattern and regional distribution of capacity availability. Given the capacity developments outlined in Chapter 4, the difficulties in meeting production targets expectedly increase over time and are worse in the West than in the East.

At operating rates below 80 percent, the 1986 steelmaking capacity of U.S. integrated producers was enough to fulfil slab demand. However, individual steel producers such as National, Sharon, Cyclops, McLouth, and some of the major steel producers, individual plants with lesser steelmaking capacity have slab deficits even at 70 percent capacity utilization (deficit here is shorthand for slab-producing capacity less than desired mill consumption of slab). In 1987 and 1988, operating rates rose above the critical 80 percent level.

Therefore, the deficiency of steelmaking capacity became evident to the point that short supply requests (allowing slab imports above the semifinished import quota) were filled for slabs (see Chapter 7).

Another significant factor contributing to the U.S. slab deficit resulted from the emergence of non-integrated flat-rolled product producers. The slab deficit of non-integrated producers rose as CSI and Tuscaloosa became operational in 1984 and 1986, respectively. Neither producer has steelmelting facilities. The Tuscaloosa plant was planned from the start not to have melting facilities. The influence of the CSI slab deficit is evident by

Table 5.10. Excess [a] of Steelmaking Capacity Over U.S. Flat-Rolled Product Finishing Capacity at Different Operating Rates--1979, 1980, 1983, 1984, and 1986 (million metric tons except as noted)

	1979	1980	1983	1984	1986
Flat-Rolled Capacity Operating Rate at 90 percent [b]					
East	0.20	1.83	1.59	-2.49	-4.90
West	1.49	0.52	-1.36	-0.99	-2.30
Total	1.69	2.35	0.23	-3.48	-7.20
Eastern Dependency [c] (percent)	0.30	2.70	2.40	-3.80	-7.60
Western Dependency [c] (percent)	43.10	14.50	-35.90	-44.80	-65.70
Flat-Rolled Capacity Operating Rate at 85 percent [b]					
East	4.60	6.04	5.68	1.44	-1.06
West	1.73	0.77	-1.10	-0.85	-2.09
Total	6.34	6.81	4.58	0.59	-3.15
Eastern Dependency [c] (percent)	6.40	8.70	8.50	2.20	-1.60
Western Dependency [c] (percent)	50.10	21.50	-29.00	-38.60	-59.80
Flat-Rolled Capacity Operating Rate at 80 percent [b]					
East	9.00	10.24	9.77	5.36	2.78
West	1.98	1.02	-0.84	-0.71	-1.89
Total	10.98	11.27	8.93	4.65	0.90
Eastern Dependency [c] (percent)	12.40	14.80	14.60	8.10	4.30
Western Dependency [c] (percent)	57.20	28.50	-22.20	-32.30	-53.90
Flat-Rolled Capacity Operating Rate at 75 percent [b]					
East	13.40	14.45	13.86	9.29	6.62
West	2.22	1.28	-0.58	-0.57	-1.68
Total	15.62	15.73	13.28	8.72	4.94
Eastern Dependency [c] (percent)	18.50	20.90	20.70	14.10	10.20
Western Dependency [c] (percent)	64.20	35.40	-15.30	-26.00	-48.00
Flat-Rolled Capacity Operating Rate at 70 percent [b]					
East	17.80	18.66	17.95	13.22	10.47
West	2.47	1.53	-0.32	-0.43	-1.47
Total	20.26	20.18	17.63	12.78	8.99
Eastern Dependency [c] (percent)	24.60	27.00	26.70	20.00	16.10
Western Dependency [c] (percent)	71.30	42.40	-8.40	-19.80	-42.10

Notes: A positive entry means an excess of steel-making capacity and a negative entry means a lack of steel-making capacity.
[a] Includes all plants of the integrated producers.
[b] Steel-making operating rate at 90 percent.
[c] The ratio of excess or lack of steel-making capacity to flat-rolled capacity.
Source: Results of the Capacity Balance Equation.

looking at the Western results. Given the region's loss of steelmaking capacity, the Western slab deficit exists whatever the operating rate.

The 1987 total semi-finished steel deficit for all grades was calculated (on the same basis of Table 5.10) using confidential data from another source. The total U.S. slab deficit amounted to 8 million tons. The slab deficit of integrated producers was estimated at 5 million tons per year, specialty

Table 5.11. Excess [a] of U.S. Steelmaking Capacity Over U.S. Flat-Rolled Product Finishing Capacity at Different Operating Rates--1979, 1980, 1983, 1984, and 1986 (million metric tons except as noted)

	1979	1980	1983	1984	1986
Flat-Rolled Capacity Operating Rate at 90 percent [b]					
East	-4.46	-1.82	-3.61	-3.67	-6.40
West	0.91	-0.07	-1.63	-1.16	-2.43
Total	-3.54	-1.89	-5.24	-4.84	-8.83
Eastern Dependency [c] (percent)	-6.20	-2.60	-5.40	-5.60	-9.90
Western Dependency [c] (percent)	26.40	-2.00	-42.90	-52.90	-69.40
Flat-Rolled Capacity Operating Rate at 85 percent [b]					
East	-0.99	1.48	-0.34	-0.44	-3.27
West	1.01	0.03	-1.53	-1.07	-2.33
Total	0.02	1.51	-1.87	-1.51	-5.60
Eastern Dependency [c] (percent)	-1.40	2.10	-0.50	-0.70	-5.00
Western Dependency [c] (percent)	29.30	0.70	-40.30	-48.40	-66.60
Flat-Rolled Capacity Operating Rate at 80 percent [b]					
East	2.47	4.78	2.93	2.79	-0.13
West	1.11	0.13	-1.43	-0.97	-2.23
Total	3.58	4.91	1.50	1.82	-2.37
Eastern Dependency [c] (percent)	3.40	6.90	4.40	4.20	-0.20
Western Dependency [c] (percent)	32.10	3.50	-37.70	-43.90	-63.80
Flat-Rolled Capacity Operating Rate at 75 percent [b]					
East	5.94	8.08	6.20	6.02	3.00
West	1.21	0.22	-1.33	-0.87	-2.14
Total	7.15	8.31	4.87	5.15	0.86
Eastern Dependency [c] (percent)	8.20	11.70	9.20	9.10	4.60
Western Dependency [c] (percent)	35.00	6.20	-35.10	-39.50	-61.00
Flat-Rolled Capacity Operating Rate at 70 percent [b]					
East	9.40	11.38	9.47	9.25	6.13
West	1.31	0.32	-1.24	-0.77	-2.04
Total	10.71	11.70	8.24	8.48	4.09
Eastern Dependency [c] (percent)	13.00	16.50	14.10	14.00	9.50
Western Dependency [c] (percent)	37.80	8.90	-32.50	-35.00	-58.20
Six Largest Integrated Companies					
At 90 percent	-3.75	-0.72	-3.41	-3.63	-5.57
At 85 percent	-0.73	2.13	-0.59	-0.84	-2.89
At 80 percent	2.29	4.99	2.23	1.94	-0.20
At 75 percent	5.31	7.84	5.06	4.73	2.49
At 70 percent	8.33	10.69	7.88	7.51	5.17

Notes: A positive entry means an excess of steel-making capacity and a negative entry means a lack of steel-making capacity.

[a] Includes only the plants producing flat-rolled products of the integrated producers.

[b] Steel-making operating rate at 90 percent.

[c] The ratio of excess or lack of steel-making capacity to flat-rolled capacity.

Source: Results of the Capacity Balance Equation.

producers at 1 million tons per year, with the non-integrated producers accounting for the rest.

Total 1987 U.S. slab trade amounted to 3 million tons. Domestic suppliers provided almost 50 percent of the needs of slab, and slab imports supplied the difference. Short supply requests accounted for 21 percent of the imported slab supplies because of the VER's total semi-finished limit of 1.8 million tons (U.S. International Trade Commission, 1988). Total 1987 U.S. hot-rolled coil trade among companies amounted to 3 million tons per year. Therefore, the total slab deficit was balanced out by the trade of slab and hot-rolled coil for re-rolling.

Increasing adoption of continuous casting by integrated producers reduced the slab deficit by one million tons. Continuous casting usage by integrated producers could almost eliminate the slab deficit at the present steel-making capacity. However, further abandonment of iron-making and associated furnaces while maintaining rolling mills would aggravate the slab supply problem.

Various firms or plants already consist of rolling mills without steelmaking capacity. CSI, Tuscaloosa, Lone Star, Heidtman, USS-Posco (a joint venture to finish Korean steel in a USS plan, [see Appendix A]), and LTV's Hennepin are examples. The impact of Heidtman, USS-Posco, and LTV's Hennepin has not been included in the calculations. They presently obtain hot-rolled wide coil for re-rolling from other plants. That coil output is already included in the calculations. Together they are expected to purchase a maximum of 3 million tons per year. Heidtman and USS-Posco, however, planned to install hot-strip mills and purchase slabs (McManus, 1986; U.S. International Trade Commission,1988). Therefore, the growth of steel producers who rely on their ability to purchase slab will stimulate the demand for slabs.

The adoption of continuous casting and the closure of non-flat-rolled product facilities made integrated producers such as USS and Bethlehem, the two largest domestic firms, potential suppliers of semi-finished shapes. USS and Bethlehem have the potential to supply 3 million tons of semi-finished steel, of which USS accounts for the large majority. My capacity-production balance calculations indicate that the Gary and Fairfield plants of USS and the Sparrows Point plant owned by Bethlehem are the largest domestic suppliers, accounting for most of the U.S. domestic supply of semi-finished steel.

If integrated mills continue to abandon certain product markets (such as bars and light structurals), leaving a plant with excess melting and casting capability, sale of semi-finished steel is one way that furnace utilization can be maintained. My computations for 1987 indicated that the closures of non-flat-rolled facilities by integrated producers made them potential suppliers of blooms and billets of 1.7 million tons per year, which reduces the U.S. bloom and billet deficit to 2 million tons.

In sum, the total U.S. semi-finished steel shapes deficiency of steel-making capacity is estimated at 10 million tons per year for capacity utilization of 90 percent. Slabs account for 8 million tons of the deficit; billets and blooms, 2 million tons. The slab steel deficit stimulated the development of new sources of domestic supplies. Thus, other domestic sources of slab are expected to increase as entrepreneurs are becoming interested in building new facilities using EF furnaces or buying old retired facilities to produce slab.

In 1988, a new slab works to be located in Arkansas with a capacity of 0.6 million to 0.7 million tons per year using a 150-tons EF furnace and continuous caster was announced. The plant will mainly produce slab for plate production which does not have as much restriction on scrap impurities as sheet. The cost of the mill was estimated at $70 million, but the plant construction is dependent on finding 4 or 5 potential investors who would buy the plant's output. These investors are expected to put up 50 percent of the amount required (*Metal Bulletin*, 1988).

In another development, the Steel Valley Authority, a quasi-governmental agency based in the Pittsburgh area, considered buying the idled South Side Works of the LTV Corporation (Pittsburgh, PA) to operate as a slab producer. The Authority reportedly has three steel companies (Cyclops, Sharon, and Lukens) interested in buying carbon steel slabs from the facility. Projected initial production was set at one million tons per year, but existing furnace capacity would allow production up to levels of about two million tons per year (U.S. International Trade Commission, 1988). Other announcements, by Michael Wilkinson, a former partner in CSI, were made concerning the purchase of Wheeling Pittsburgh's Monessen facilities (closed in 1987) to produce slab.

OTHER FACTORS CONTRIBUTING TO SLAB IMPORTS

Other factors besides the difference between steelmaking and finishing capacity affect the supply and demand of slabs. A key aspect of the slab deficit is differences between the East and the West. Tables 5.10 and 5.11 show the trends of this regional deficiency of steelmaking capacity from 1977 to 1986. From 1977 to 1986, the Western slab deficiency of steelmaking capacity changed from a surplus to a deficit of 2 million tons. The Eastern slab deficit grew, reaching a deficit of 5 million to 6 million tons in 1986.

Western dependence on slab was greater than that in the East, as mentioned above. However, the Western dependence on slab can be measured differently by using the ratio of the slab deficit to flat-rolled capacity. Both regions demonstrate an increasing reliance on slab imports, but since 1982, Western employment has been consistently much higher than Eastern employment. To maintain 90 percent utilization of 1986 Western finishing capacity, slab imports would have had to supply 66 percent to 69 percent of the slab needed. The corresponding Eastern import share was 8 percent to 10 percent.

The greater relative dependence in the West on slab imports resulted from the closure of hot-end facilities. Eastern reliance on slab imports disappears when the finishing capacity utilization is at or below 75 percent to 80 percent even if only the steelmaking capacity of plants producing flat-rolled products is considered.

Inland freight rates discourage Western producers from purchasing slabs from the East. Rail transportation costs range from $38 to $83 per ton, depending on plant location. Shipments from the East to the West could dramatically reduce import dependence in the West, if lower freight rates could be attained through ocean shipping. However, the Jones Act requiring use of United States registered ships in interstate commerce precludes attaining such a cost reduction.

Other factors affecting the slab deficit are episodic demands caused by closure for the maintenance or rebuilding of hot-end facilities and the deficiency of specific types capacity within the hot-end facilities. Required maintenance on furnaces and casters (ranging from 4 months to 8 months) may lead producers to seek slab to cover shortfalls until such facilities are brought back on line.

Table 5.12 shows the discrepancy, from 1977 to 1985, between coke oven and blast furnace capacities at different blast-furnace capacity utilization rates, assuming a coke rate of 0.5 million tons of coke to produce 1 ton of pig iron (data for 1986 are not available). A coke deficit occurred when high blast-furnace operating rates prevailed and a surplus of coke exists when blast furnaces operate at 70 percent or below. Low blast-furnace capacity utilization hid the potential coke deficit that occurs when blast-furnace production rises. Some firms such as Weirton, Sharon, and Cyclops that are now producing steel from EF and some individual plants of the major integrated producers are entirely dependent on coke purchases.

The discrepancy between coke-oven and blast-furnace capacity has not been influential. At a wide range of possible scrap charging rates, surplus blast-furnace capacity exists (Table 5.13). In the calculations, I assumed that the BOF and OH always operated with molten pig iron and scrap, EF operated with 100 percent scrap, and the maximum possible capacity utilization rate was 90 percent for both steelmaking and blast furnace capacities. From 1977 to 1986, excess pig-iron capacity dropped from 32 million to 19 million tons for 30 percent scrap charging of BOF and OH. However, lack of investment also may reduce blast-furnace capacity (see Chapter 6).

Blast-furnace relining was delayed as long as possible by expedients noted in Chapter 2. Therefore, an estimated 28 million tons of blast-furnace capacity needed relining by 1989; given a scrap charge of 30 percent, 41 million tons of crude-steel production capacity would lack the necessary pig iron (Table 5.14). If this relining were not undertaken, a 9-million-ton deficit would exist. Conversely, 26 million tons per year of idle blast-furnace capacity equivalent could be brought into production and would alleviate the pig-iron deficit when relining of presently operating blast-furnace capacity is required. Although reactivating idle blast-furnace capacity could lessen the total U.S. pig-iron deficit, the needs (see Chapter 2) of individual producers for molten pig iron to charge BOFs may not be covered. The pig iron needs of such individual producers is estimated to generate a slab demand of anywhere from 1 million to 2 million tons.

The 1987-88 price increase of scrap reflected another aspect of the discrepancy between blast-furnace and steelmaking capacity, namely the changes in scrap-supply conditions discussed in Chapters 2 and 3. From

Table 5.12. Excess of U.S. Coking Capacity over U.S. Blast Furnace Capacity at Different Operating Rates-- 1979, 1980, 1983, 1984, and 1985 (million metric tons)

	1979	1980	1983	1984	1985
Coke and Blast Furnace Capacities Imbalance					
Blast Furnace at 90 percent	0.43	0.66	-6.85	-1.98	-1.08
Blast Furnace at 85 percent	3.24	3.27	-4.57	-0.04	0.81
Blast Furnace at 80 percent	6.05	5.87	-2.29	1.90	2.70
Blast Furnace at 75 percent	8.85	8.48	-0.02	3.84	4.59
Blast Furnace at 70 percent	11.66	11.09	2.26	5.78	6.48

Note: A positive entry means an excess of coking capacity and a negative entry means a lack of coking capacity.
Source: Results from Capacity Balance Equation.

Table 5.13. Excess Capacity of U.S. Blast Furnace Capacity Over U.S. Steelmaking[a] Capacity Both at 90 Percent Operating Rate and Different Scrap Charging--1979, 1980, 1983, 1984, and 1986 (million metric tons)

	1979	1980	1983	1984	1986
Scrap Charging at 30 percent	34.35	30.14	23.43	19.28	19.22
Scrap Charging at 25 percent	29.58	25.59	19.24	15.66	15.93
Scrap Charging at 20 percent	24.82	21.04	15.06	12.05	12.65
Scrap Charging at 15 percent	20.06	16.49	10.88	8.44	9.36
Scrap Charging at 10 percent	15.29	11.94	6.69	4.83	6.07
Scrap Charging at 5 percent	10.53	7.39	2.51	1.22	2.78
Scrap Charging at 0 percent	5.77	2.84	-1.67	-2.39	-0.50

Notes: A positive entry means an excess of blast-furnace capacity and a negative entry means a lack of blast-furnace capacity.
[a] It takes into account only basic oxygen and open-hearth furnaces because electric furnace is considered to operate with 100 percent scrap.
Source: Results from Capacity Balance Equation.

1977 to 1986, home-scrap supply declined 40 percent, to 16 million tons. As a result, the home-scrap share of the total apparent supply of scrap is estimated to have dropped from 59 percent to 44 percent, and more prompt and obsolete scrap had to make up for the difference.

However, the amount of prompt supply also was declining because of the increasing efficiency of steel users. The effect of limited home and prompt scrap supplies and increasing operating rates of the U.S. industry in 1987-88 produced a rise in the composite scrap price of 47 percent (*American Metal Market,* 1988). At the 1988 level, the cost of liquid steelmaking in integrated and mini-mills is almost equivalent, and a further decline in quality scrap supplies could raise prices to a point where it is not economical to charge scrap in the BOF or OH furnaces. If this is the case, the discrepancy between blast furnace and steelmaking facilities would result in a pig-iron deficit. A 90 percent capacity utilization of blast furnace

94 Implications of Imported Slabs in the U.S. Market

Table 5.14. Estimated U.S. Coke Oven and Blast Furnace Capacity Requiring Relining (million metric tons)

Expected Life	1 to 3 years	4 to 6 years	6 to 9 years
Coke Oven Relines			
Coke Capacity	6.61	3.89	4.14
Pig-Iron Equivalent[a]	13.21	7.77	8.28
Crude Steel Equivalent[b]	18.87	11.10	11.83
Blast Furnace Relines			
Pig-Iron Capacity	28.35	22.04	8.63
Crude Steel Equivalent[b]	40.50	31.49	12.32

Note: [a] Assumes a coke rate of 0.5 kilograms per ton of pig iron.
[b] Assumes a scrap charging of 30 percent.
Source: *33 Metal Producing*, "World Steel Industry Data Handbook--USA," 1985b.

and steelmaking and scrap charges ranging between 0 and 5 percent could produce such a deficit.

Therefore, the balance between blast-furnace and steelmaking capacity is maintained by scrap charges of at least 5 percent. Increasing scrap prices would affect the international cost competitiveness of U.S. integrated producers against countries, such as Brazil, not as dependent on scrap charges.

The shut-down of the remaining OH furnace capacity is likely to occur given their age and the cost of their maintenance. This, too, will increase the slab-producing capacity deficit. The last factor affecting the slab deficit is related to quality relating mainly to the availability of continuously cast slab. Continuously cast slab produces better yields in finishing and provides more homogeneous physical and chemical quality properties than slab produced from ingot casting. Some firms lacking the ability to secure high quality steel from continuous casting (Weirton and National) purchased such steel to make products with stringent quality requirements. In some cases, imported steel was bought. The higher quality provided by continuously cast steel was reflected by a surcharge of $20 to $40 per ton for continuously cast slab against slab produced from ingot casting. Increasing adoption of continuous casting lessens this problem.

In sum, the capacity reductions summarized in Chapter 4 resulted in a discrepancy between steelmaking capacity and finishing capacity, especially in the West. This discrepancy resulted in a slab deficit capacity that is noticeable when operating rates are above 80 percent. Slab imports could be the most economic way to overcome this deficit.

Implications of Imported Slabs in the U.S. Market

SUMMARY AND CONCLUSIONS

This chapter examined rising slab imports. Reducing integrated producers' capacity led to inability to produce the slab demanded for finishing. Imported slab proved valuable to flat-rolled producers, especially in the West where import penetration of flat-rolled products was reduced by importing slab.

6

Steel Production Costs and the Prospects for Slab-Steel Imports

Preceding chapters examine trends in the world steel industry. Particular attention is given to rising international trade in steel slabs. This chapter evaluates the available data on historic and future steel-supply economics with particular stress on the potential for increasing slab-steel imports to the United States.

However, data are presented on factors affecting costs throughout the steel industry and on the economics of non-flat-rolled as well as flat-rolled products production. Thus, the discussion also recapitulates evidence about why U.S. integrated producers lost markets to imports and domestic mini-mills.

This chapter discusses U.S. international competitiveness in producing slabs and evaluates the incentives to import slab to produce flat-rolled products. The costs and profitability of different ways of producing a mix of flat-rolled products are appraised. The alternatives considered include non-integrated plants that purchase, rather than produce, slab from domestic or foreign suppliers and new and existing integrated plants.

Considerably more data than are available are always required to conduct supply analyses, and this principle applies to steel. Therefore, discussion begins with review of critical issues in steel supply analysis. Then, the nature of available data and the procedures used to analyze them are discussed. Information is next provided about international differences in the costs of key inputs.

Discussions cover labor, iron ore, and coking coal. Prior steel production cost estimates are summarized. Alternative, more detailed cost appraisals that I developed are then presented. Both the prior estimates and my initial ones use simple standard cost-accounting methods to determine the costs. Then, a present-value approach is used to determine the comparative costs correctly.

PRINCIPLES OF SUPPLY ANALYSIS

Economic theory indicates that supply at any moment can be provided by facilities that differ considerably in their operating costs and that the situation will change over time. Inaction will cause facilities to deteriorate, raise costs, and probably lower the physical ability to produce. This deterioration can be offset by investments in the rehabilitation and in the expansion of existing plants and the construction of new ones.

Further changes occur over time in the productivity which are possible with available technology and in the costs of inputs to production. With an internationally traded commodity in a world in which major changes in real exchange rates often occur, adjustments to changes in exchange rates must be made. For many commodities including steel, transportation costs constitute a significant enough part of the delivered price that the location of sellers vis-à-vis consumers is a major influence on the ability to compete. The analysis of markets for such goods must explicitly consider the locations of suppliers and markets and the costs of transportation between each supplier and each consuming location.

This would require constructing an analysis far beyond the scope of a single book. At a minimum, a cost analysis of every existing plant, detailed data on the location of customers, and the required transportation cost data would have to be included. To resolve completely the questions treated in this book, much more would be needed. Enough information should be provided on the age and condition of the equipment in each plant and on the costs of equipment replacement to permit analysis of the economics of maintaining capacity. Appraisal should be provided of the extent to which the marginal cost of production will rise as the output of any given steel industry increases. Finally, the analysis should consider alternative outcomes from as yet unproved new steelmaking technologies.

Only indicators of the possibilities are available. The discussions in prior chapters suggest some of the issues that must be considered. First, the dispersion of the age and location of steel-mill facilities is considerable. Thus, differences prevail among plants within any country in the current cost of production and the costs of maintaining that capacity.

However, the evidence suggests that technological factors make portions of any integrated steel mill particularly vulnerable to deterioration. In particular, regular investment is required to maintain blast-furnace capacity. The techniques noted in Chapter 2 for delaying full relining can cause deterioration in the facility that increases the cost of full rehabilitation when it occurs. The present value analysis here includes the costs of blast furnace rebuilding at least once every six years.

Another concern is that of the changing comparative economics of production from scrap as steel-supply evolution proceeds. Chapter 2 indicates that the rise of continuous casting and the improved technologies in steel-consuming industries reduce the supply of home and prompt-industrial scrap. Thus, more reliance must be placed on the more expensive-to-use obsolete scrap. For any given scrap-supply situation, the advantage of scrap use diminishes as the proportion of steel made from scrap rises.

Limits on the ability of the BOF to accept scrap, and more critically the possibly prohibitive costs of meeting quality specifications using large proportions of scrap, preclude total elimination of production of steel from pig iron. New technologies could reduce some of these disadvantages of scrap use.

On the foreign-supply side, a critical question is the extent to which suppliers such as South Korea and Brazil can expand slab-steel output without facing significantly increasing costs. Another issue is whether the economics of direct reduction iron (DRI) would allow increased production and export abroad. This depends upon both how successful a new process such as Corex proves to be, and also on the extent to which low cost energy, such as the presently underutilized natural gas resources in some oil-rich countries, remains available.

Some analysts of DRI believe that American steel producers can purchase merchant DRI from Latin America at costs lower than those for U.S. pig-iron production (Innace, 1985a). Such imports might be particularly attractive as a supplement to scrap for mini-mills. Rising prices as scrap supplies tighten and rising processing costs as scrap quality decreases might make DRI more competitive in mini-mills.

As a result, DRI proponents anticipate and are keying their strategies toward expanding the market for merchant DRI in the United States. U.S. steel producers import on an average only 100,000 to 200,000 tons of merchant DRI per year for EF charging. DRI proponents estimate that the United States has the potential to consume 2 million tons of merchant DRI, which would represent 10 percent to 30 percent of a furnace charge (Innace, 1985). Increased import of merchant DRI might facilitate production of flat-rolled steel in mini-mills by providing a higher quality steel. Merchant DRI would compete with quality scrap because the delivered prices of merchant DRI from Latin America range from $80 to $130 per ton (*World Steel Dynamics*, 1987).

Data on the veracity of these contentions were not obtained. At the scales envisioned, DRI imports would have trivial impacts. Any threat to either present patterns or the increased use of slabs stressed here would require an ability to use DRI to produce large quantities of steel usable for flat-rolling at competitive cost.

Several other outcomes might emerge from the economic success of DRI, thin slab casting, and other technologies. An entirely different type of integrated mill located in the big market countries might become highly competitive. Such a new type of mill might use a higher proportion of scrap and operate on a smaller scale than does the traditional integrated mill. The technology might alternatively allow both new and existing producers of different sizes to remain competitive, revitalize the position of the large mill, or favor smaller producers. Conversely, new technology might allow much greater reliance on scrap-based steelmaking.

The present discussion largely assumes that DRI trade is unlikely to become significant and that slab made in the BOF from pig iron and scrap will continue to be the cheapest input to production of flat-rolled products. The assumption about DRI seems realistic because of the limits to cheap energy supplies. Also, merchant DRI and pig iron have an intrinsic loss of metal content in shipping because of the degradation during loading and unloading that raises the cost of delivering such a product. This problem may be overcome with the development of new iron-making technologies or briquetting of DRI.

Therefore, the analysis stresses cases in which some form of BOF-based slab production remains critical and applies to increases of slab imports that can be met by the amount of capacity Chapter 3 showed would be available. However, estimates of thin slab-casting costs are considered.

It is further assumed, again to simplify the analysis, that prices of raw materials, plant sites, and other inputs will not rise significantly as steel output increases. Therefore, the price of slab imported from countries such as Brazil and South Korea will not change if U.S. producers increase their purchases to the extent considered here. Most steel industry people and steel analysts consulted believed that expansion in Brazil of the magnitude considered here was possible without increasing marginal costs. Given improvements in steel industry technology and the management experience gained by Siderbras, new plants might be cheaper to build and operate than existing ones.

The analysis does assess the sensitivity of the results to slab prices, exchange rates, and the location and efficiency of existing plants. In particular, comparisons are made of the economics of different ways of securing slab at new facilities to replace capacity now or shortly to be abandoned, existing plants and newer plants on the West Coast, the East Coast, and an inland location such as Chicago.

Another problem in supply analysis is that what accountants consider sound practice involves bad economics. The defects include failure to adjust for price-level changes, unsatisfactory treatment of investment outlays, and arbitrary allocation of the common costs of an integrated facility. Under some circumstances, accounting data can be adjusted to provide reasonable estimates of costs or at least the differences among costs of different supply alternatives. Costs can be adjusted to a constant dollar basis, and an estimate of the required "levelized" profits can be added.

Levelizing is a widely used convention of measuring capital costs as the annual income that would be earned at the cost of capital if the investment were amortized in equal annual installments. In this book, the common cost allocation is eliminated by considering the total cost of producing the entire product mix of a plant producing 4 million tons of flat-rolled finished products.

A more serious problem is that locational differences among firms are not considered in most published works. In this book, the locational differences are treated by considering shipping costs from U.S. and foreign slab producers to U.S. slab purchasers and from finishing mills to final consumers.

Given the widespread tendency to think in per-ton terms, the totals are divided by output to produce a weighted average. Since this involves dividing by a constant, differences in comparative weighted average costs imply a similar comparative difference in total costs. This analysis is designed to compare the costs of alternative routes to the producing of flat-rolled products. The approach can be applied to new plants by viewing total costs or to existing plants by viewing operating costs. Appendix D outlines the costing method used.

This technique is often used by steel producers, consultants, governments, and steel industry analysts, such as those cited in this chapter, to estimate the average cost of producing steel in different countries. This procedure does not measure the marginal cost of producing steel that is the foundation of the Marshallian supply curve. However, the method is useful for epitomizing international cost competitiveness and suggesting the profitability of different production strategies.

Finally, true present value analyses are undertaken under various assumptions about the investment outlays required to maintain different types of operations and flat-rolled producer perceptions of future scenarios. The choice among traditional pig-iron-based steelmaking methods, DRI-based steelmaking, and slab imports is stressed here.

AN OVERVIEW OF COMPARATIVE INPUT SUPPLY PROBLEMS OF STEEL PRODUCERS IN THE UNITED STATES AND ABROAD

As the pro forma cost estimates presented below indicate, labor and iron ore are the main sources of cost differences between U.S. and foreign integrated producers. The U.S. is a higher cost coking-coal producer than Australia and Canada, the main suppliers of Japan and South Korea. However, U.S. transportation-cost advantages offset production-cost disadvantages and give U.S. steel producers lower coke costs than their foreign competitors.

The EEC coking-coal situation is distorted by protection of high-cost domestic production. At one extreme, German coal producers have assured markets with subsidies provided to cover the difference between imported coal prices and German production costs. Other countries are phasing out protection and relying on imports (Gordon, 1987; IEA, 1988a and 1988b). No major capital-cost advantages appear to exist among countries.

As noted, mini-mills have an advantage over domestic and foreign integrated producers resulting from their ability to produce non-flat-rolled products entirely from steel made from scrap melted in an electric furnace. At prevailing U.S. scrap prices, scrap is less costly than pig iron, and in the products produced by mini-mills, scrap is a nearly perfect substitute for pig iron. Two further advantages of the mini-mills are their greater proximity to growing markets than the integrated mills and their use of non-union labor.

Thus, what is most critical to examine are the real problems of high labor costs in the United States and the high reported levels of iron-ore costs. The labor problem arises from strong commitments made over the years to the United Steel Workers (USW). The nature of the iron-ore problem is less evident and needs exploration.

Obligations to the USW and the costs of lay-offs hinder adjusting payments to workers to changing U.S. labor market conditions. Productivity improvement is much more feasible than changing worker compensation. Labor costs per ton increased consistently until 1981. Since

1983, they decreased due to the re-negotiation of benefits, working rules and, most importantly, increasing labor productivity. Unit labor costs fell substantially in 1983 and 1984 as the troubled integrated companies closed facilities, reduced work crews, and shed salaried personnel.

Steel workers traditionally earned total wages and fringe benefits above the average for American manufacturing workers. From 1952 to 1962, this premium rose from 18 percent to 36 percent. In 1968, intensifying foreign competition caused the premium to drop to less than 30 percent.

In 1973, it increased to 40 percent when the demand for steel products was the strongest in the entire history of the American steel market; high demand also prevailed in other steel markets of the noncommunist world. Forecasts by analysts and industry executives in the United States predicted tight demand well into the 1980s. This caused concern about a strike and its impact on the market. Thus, the industry signed the Experimental Negotiating Agreement (ENA) with steel workers that secured bans on nationwide strikes by making substantial wage concessions.

U.S. integrated steel producers soon recognized that the ENA greatly increased production costs (U.S. Congressional Budget Office, 1984). Hourly employment costs for steel workers rose so that by 1981 they exceeded the manufacturing average by 73 percent. By 1982, the premium reached 92 percent. A deep recession and import competition forced the industry to seek adjustment. After lengthy negotiation and two refusals, the union finally voted for concessions that put the premium back approximately to its 1981 level, i.e., far above its pre-protection level of less than 30 percent.

After 1985, total U.S. employment costs per hour began increasing again as a result of heated labor-management negotiations and increasing crude-steel production. Wage concessions did not suffice to restore the competitive position of the U.S. integrated sector.

In the equally troubled European steel industries, total compensation is generally only 15-20 percent above the manufacturing average (Barnett and Crandall, 1986).

Much attention has been paid to the impact of pension obligations on plant closings. These pensions are intended to be fully funded by employer contributions made while the workers were employed. However, employers' contributions to pension plans largely are based on previous experience. Therefore, the pension funds' resources are often inadequate to cover obligations created by terminations that exceed the historical rate, and the firm must make additional payments to keep the pension fund fully funded.

In the U.S. steel industry, such obligations to laid-off workers apparently have affected operating decisions. As with any commitment to severance costs, the pension obligations reduce the benefits to the firm of layoffs. The reduction of benefits of job elimination leads to fewer dismissals than would otherwise occur. These costs are the unexpected costs of past production rather than the costs of future production. Thus, an inefficient excess of private costs over social costs has been created.

Nevertheless, declining steel shipments, the shift to lower cost techniques, changes in operating practices (such as working rules), and the elimination of many salaried positions reduced employment in the steel industry (Table 6.1). Even though steel production increased 24 percent between 1982 and 1984, employment continued to drop. In the past, such an increase in production would have boosted industry employment. These changes are likely to be permanent.

Iron Ore

Available data usually employ the prices of iron ore reported by the companies. This probably does not accurately reflect the true costs since these prices are, to a considerable extent, transfer prices used by vertically integrated iron-ore mining operations.

From 1973 to 1986, the reported weighted average U.S. cost of iron ore from domestic and foreign sources increased from $18 to $42 per ton. As a result, in 1986 the apparent U.S. iron-ore price was far above that of other net importers of iron ore. Compared to Japan, the U.S. iron-ore price was $16 per ton higher (*World Steel Dynamics*, 1987).

As Chapter 3 indicates, while other the big-market countries have increased their reliance on imported iron ore, the United States has not. In particular, the U.S. has not turned to Brazilian or Australian ore to the extent that other countries have.

The most plausible explanations for these patterns are the location of U.S. mills and the high degree of vertical integration prevailing between steel companies. Steel companies in the Great Lakes market account for 75 percent of total U.S. iron-ore consumption; they produce it largely from their own mines in the United States. Thus, the marginal variable costs of these mines are more relevant than reported transfer prices. The variable costs of North America's ironore are about $2 per ton more expensive than that of imported iron ore. The locational advantage of the domestic and Canadian suppliers may suffice to justify such a premium. Changing ore suppliers, therefore, is not as economic as the price data suggest (Franz, 1986).

While this argument explains such behavior, it also implies that prior writers have overstated U.S. costs by overstating the true cost of iron ore. Here an effort is made to adjust the data to provide more realistic figures. In particular, the ore cost is lowered to that of imported ores sold by independent suppliers.

Locational Factors

The geographical dimension of the market is a further influence because freight costs may comprise a large portion of the delivered cost. What determines the competitiveness of a supplier in a region is not the f.o.b. price, but its delivered cost, adjusted for product quality. The impacts of transportation costs naturally differ with customer locations. The further inland and the closer a customer is to existing steel mills, the greater is the freight cost advantage of domestic producers.

Table 6.1. Number of Employees and Total Employment Costs per Hour Worked in the U.S. Steel Industry --1973 to 1987

Year	Number of Employees (thousands)	Total Employment (hourly employees) Dollars per Hour Worked
1973	509	7.7
1974	512	9.1
1975	457	10.6
1976	454	11.7
1977	452	13.0
1978	449	14.3
1979	453	15.9
1980	399	18.5
1981	391	20.2
1982	289	23.8
1983	243	22.2
1984	236	21.3
1985	208	22.8
1986	175	23.2
1987	163	23.7

Source: American Iron and Steel Institute, *Annual Statistical Report*, various years.

Conversely, the closer a customer is to the East or West Coast and the further the distance from a domestic steel mill, the greater is the attractiveness of imports. For present purposes, however, the location of U.S. steel mills is the critical difference.

Freight cost varies, not only with the distance between seller to purchaser, but also with the size of shipment, return cargo possibilities, and demand for transport. In the case of ocean freight, cost further varies with the size and flag of the vessel. Insurance costs vary with the nature of the product, record of the seller, record of unloading, and loading locations.

Historically, ocean-freight rates were less than inland railroad rates for raw materials and steel. Before U.S. government deregulation of train and truck freight, a ton of hot-rolled coil produced in Chicago would cost at least an additional $100 per ton to ship to Los Angeles by train (Mueller, 1982). The same product could be shipped from coastal steel plants in Japan, Europe, or Latin America to the United States for only $55 to $65 per ton including tariffs and insurance (Mueller, 1982).

Thus, transportation-cost advantages reinforced production-cost advantages of foreign suppliers in U.S. coastal markets. In 1987, Japan placed 43 percent of its exports to the United States of plate and hot-rolled and cold-rolled sheets in the West and 57 percent in the East, of which only 7 percent were imported to the Great Lakes.

After U.S. railroad deregulation, inland rates fell to $45 per ton. The sufficiency of this reduction is unclear. Not only is it not necessarily sufficient to offset higher production costs, but ocean freight rates also have declined (Table 6.2).

In contrast, the average of customs and other costs for exporters to the United States increased from 1973 to 1986 (Table 6.2). From 1981 to 1986, the total shipping cost for U.S. importers decreased from $98 to $76 per ton. After 1986, ocean-freight rates declined further and, depending on the exporting country and flag of the vessel, freight rates to the United States could be as low as $25-$30 per ton for finished steel. In the case of semi-finished steel, the rate can be as low as $15-$20 per ton because shipments are larger than those of finished steel.

Exchange Rate Effects

Changes in exchange rates greatly affect competitiveness in steel. Exchange rate changes can arise from differential inflation rates, changing real competitive position, or both. To the extent real forces are at work, trade patterns will alter. The real devaluation of a currency lowers the cost in a foreign currency of steel output and raises domestic currency imported steel costs. Therefore, real depreciation of a currency stimulates exports and reduces imports. However, the effect of exchange rate movements is not straightforward since some steelmaking costs, such as for coal and iron ore, are denominated in dollars and some exporters, such as Brazil and South Korea, try to maintain a stable exchange rate against the dollar.

To indicate major changes in exchange rates, the indexed real exchange rate that takes into account the inflation of the United States and foreign countries was calculated (Tables 6.3 and 6.4). The U.S. dollar depreciated against major currencies, especially Japan and for countries in the EEC. The dollar devaluation against Brazil, South Korea, and Canada was not as dramatic as against the other countries since Brazilian and South Korean currencies were closely pegged to the U.S. dollar. The U.S./Canadian dollar exchange rate did not change as sharply as did the relations of the U.S. dollar to the mark and to the yen. These last two changes were critical influences on international competitiveness.

AN OVERVIEW OF STEEL COMPARATIVE ECONOMICS

Barnett and Crandall's (1986) review of the rise of the mini-mill presented estimates of steel production costs. Barnett has regularly updated some of these data. Given the contribution these estimates make to understanding steel supply trends and the extent to which my methodology is related to that used in these prior cost estimates, a review seems appropriate.

First, two basic cost comparisons made by Barnett and Crandall are presented. One used wire-rod production cost comparisons to illustrate how well a U.S. mini-mill could compete with integrated producers in the U.S.

Table 6.2. Contract and Spot Ocean Freight Rates and Customs and Other Costs Associated with Delivering Steel Products to the United States--1973 to 1986 (dollars per metric ton)

Year	Ocean Freight Rates		Customs and Other Costs
	Contract	Spot	
1973	36.5	28.3	19.5
1974	45.7	44.3	30.1
1975	35.3	40.5	28.2
1976	31.6	32.2	26.2
1977	30.5	31.3	27.5
1978	34.4	32.0	33.3
1979	49.2	41.4	39.2
1980	57.2	55.2	41.8
1981	50.9	55.1	45.0
1982	39.7	44.5	42.7
1983	38.5	38.1	39.7
1984	38.3	38.3	40.9
1985	38.2	38.6	40.7
1986	32.9	35.4	41.6

Source: *World Steel Dynamics*, "Steel Strategist #14," 1987.

and Japan. The other employed comparisons of cold-rolled sheet costs to show the comparative position of different integrated producers.

The works by Barnett stressed here are his updatings of the cold-rolled sheet cost data and his effort to show the variation of costs among U.S. mills. He presented estimates of how costs differ with the quality of the plant and of how much capacity was available in different quality categories.

Competitiveness of U.S. Integrated Producers in Non-Flat-Rolled Products--The Case of Wire Rods

Estimates of the costs of producing wire rods are available for integrated and mini-mill producers. From 1984-85 to 1986-87, mini-mill capacity in this sector increased from 39.4 percent to 53.8 percent of total U.S. capacity. The cost of producing wire takes into account the input factors, quantity, and price, the yields of each production stage, and the indirect costs allocated on a production basis.

Table 6.5 presents estimated 1985 costs of producing wire rods for a representative U.S. mini-mill and integrated producers in the United States and Japan. Mini-mill costs were $35 per ton below those of integrated Japanese producers and more than $100 per ton below those of U.S. integrated producers. In 1987, by adopting new strategies, the operating cost of producing wire rod in an integrated plant was lowered to $308 per ton shipped, an 18 percent cost reduction. The operating cost of mini-mills was $264 per ton shipped (Barnett, 1987).

Table 6.3. Nominal Exchange Rates for the Main Countries Exporting to the U.S. Market--1973, 1979, 1983, and 1987

		1973	1979	1980	1983	1987
Japan	Yen	271.2	219.2	226.7	237.5	144.6
Canada	C.Dollar	1.0	1.2	1.2	1.2	1.3
West Germany	Mark	2.7	1.8	1.8	2.6	1.8
South Korea	Won	398.3	484.0	607.4	775.8	822.6
France	Franc	4.5	4.3	4.2	7.6	6.0
Brazil	Cruzado	0.0	0.0	0.1	0.6	39.2
Netherland	Guilder	2.8	2.0	2.0	2.8	2.0
Belgium	B.Franc	39.0	29.3	32.3	51.1	37.3
United Kingdom	Pound	0.4	0.5	0.4	0.7	0.6
Sweden	Kronor	4.4	4.3	4.2	7.7	6.3
Mexico	Peso	12.5	22.8	23.0	120.1	1,378.2
Venezuela	Bolivare	4.3	4.3	4.3	4.3	14.5
Italy	Lire	583.0	830.9	856.5	1,518.8	1,296.1
South Africa	Rand	0.7	0.8	0.8	1.1	2.0

Source: International Monetary Fund, *International Financial Statistics*, Various Years.

Another way to appraise the competitiveness of integrated and mini-mill producers is to look at the trends in the import prices and quantities of non-flat-rolled products. Table 6.6 shows the import penetration and the ratio between imported and domestic shipments prices for typical products of integrated and mini-mill producers.

The import penetration of wire rods and hot-rolled bars either declined or stayed at the 1977 level. The ratio between import and domestic prices for hot-rolled bars and wire rods declined to a point at which, from 1981 to 1986, domestic prices were lower. The domestic prices of wire rods were consistently lower than imported steel since 1983, and domestic prices of hot-rolled bars improved relative to import prices during the 1979-87 period.

Although the mini-mill producers are not the sole producers of these products, their growth in the market share significantly influenced the fall of these ratios. Therefore, the competitive advantage of mini-mills over foreign suppliers reduced import penetration of non-flat-rolled products markets, and the mini-mills competitive edge over integrated domestic producers lowered the market share and capacity of integrated non-flat-rolled products producers.

The Competitiveness of U.S. Integrated Producers in the Flat-Rolled Products Market

In the flat-rolled product market, integrated producers are faced only with imported steel competition, which is restrained by the VER. However,

Table 6.4. Real Exchange Rates for the Main Countries Exporting to the U.S. Market --1973, 1979, 1983, and 1987 (1977 base)

		1973	1979	1980	1983	1987
Japan	Yen	101.5	94.7	94.9	111.0	79.6
Canada	C.Dollar	98.0	106.9	107.3	105.5	104.3
West Germany	Mark	101.0	90.3	94.9	129.8	92.7
South Korea	Won	125.5	91.5	94.3	107.4	115.1
France	Franc	86.9	96.1	92.6	137.4	n.a.
Brazil	Cruzado	100.8	107.9	117.3	133.8	113.2
Netherland	Guilder	105.5	95.3	99.6	140.4	99.1
Belgium	B.Franc	97.8	95.2	112.9	164.5	124.6
United Kingdom	Pound	104.0	81.6	72.9	101.3	78.4
Sweden	Kronor	107.3	95.9	94.8	139.3	102.9
Mexico	Peso	89.7	89.5	82.5	120.8	121.3
Venezuela	Bolivare	109.1	103.5	98.3	84.3	125.4
Italy	Lire	101.0	91.3	89.3	121.7	87.5
South Africa	Rand	100.1	93.1	84.1	94.7	102.1

Source: International Monetary Fund, *International Financial Statistics*, Various Years.

integrated capacity differs considerably in production cost and product quality.

Tables 6.7, 6.8, and 6.9 present the cost of producing cold-rolled sheet by efficient integrated steel producers in the United States, West Germany, Japan, South Korea, and Brazil, in 1985, 1986, and 1987, respectively. The data for 1985 assume 90 percent capacity utilization for all countries. The 1986 and 1987 cost estimates are for the actual capacity utilization rates prevailing in each country in each year. These estimates indicate that, although the United States had higher labor productivity than the other countries except Japan, high wages were the major source of the competitive disadvantage of U.S. integrated producers.

The data also indicate high iron-ore costs, but, as noted above, questions about the relevance of reported iron-ore costs make these data suspect. The more realistic figures shown in Tables 6.14 and 6.15 suggest that relative cost positions are not significantly altered by using different figures for iron-ore prices.

Another cost-raising factor was a lower yield of finished products due to the lower rate of adoption of continuous casting technology. Both the operating and total cost of producing cold-rolled sheet in the United States were higher than for the other four countries in 1985 and 1986.

Tables 6.8 and 6.9 also show the effects of the 1986-87 devaluation of the dollar against major currencies. In 1987, the ratio of import and domestic prices for all the products improved compared to 1985 when the dollar started to devalue. The figures for 1986, as noted, are those at the capacity utilization prevailing in each country. All calculated per-ton fixed costs, here comprising both those explicitly termed fixed costs and some miscellaneous costs, necessarily will be higher if the operating rates are

Table 6.5. Cost of Producing Wire Rod at Representative U.S. Mini-Mills and Integrated Producers in the U.S. and Japan in 1985 (dollars per metric ton of finished product)

	U.S. Mini-mill	U.S. Integrated	Japan Integrated
Iron Ore		67	49
Scrap	105	15	2
Coal and Coke		46	65
Other Energy	49	22	12
Total Raw Materials and Other Energy Costs	154	150	128
Labor	46	123	52
Miscellaneous	69	99	104
Operating Costs	269	374	283
Depreciation	10	12	19
Interest	13	9	19
Taxes	2	4	4
Total Cost	294	399	325
Input Prices			
Labor (dollars per man-hour)	17.5	22.3	11.7
Iron Ore (dollars per ton)		44.1	26.7
Scrap (dollars per ton)	93.7	88.2	96.5
Coal (dollars per ton)		60.6	65.6
Electricity (cents per kilowatt hour)	4.5	4.5	7.0
Yen per dollar			240
Efficiency Measures			
Man-hour per ton	2.7	5.5	4.4
Yield to Finished Product (percent)	94.0	88.0	93.0

Source: Barnett and Crandall, *Up from the Ashes*, 1986.

lower. This was true for U.S. production costs since a 65 percent capacity utilization rate prevailed.

Estimated total labor compensation and iron-ore prices declined, reducing both labor and iron costs by $4 per ton each. Also, a higher finished product yield was attained as integrated producers increased their continuous- casting capacity.

The cost competitiveness of U.S. producers improved against West Germany and Japan because of the dollar devaluation. The dollar labor, depreciation, and interest costs of West Germany and Japan rose. Domestic currency costs fell but not enough to offset the dollar depreciation. Meanwhile, the cost competitiveness of the American with Brazilian and South Korean producers remained unchanged because those countries have their currencies pegged to the dollar.

In 1987, the total costs of producing cold-rolled sheets by U.S. integrated producers fell below those of West German and Japanese producers. This was largely due to the further devaluation of the dollar. The impact is noticeable when looking at dollar labor costs, which increased

Table 6.6. U.S. Import Penetration[a] and Ratio of Imported and Domestic Prices for Selected Products--1977 to 1987

	1977	1979	1980	1983	1987	Percent Change 1977-1987
Plate						
Import Penetration	0.4	0.3	0.3	0.5	0.5	39.9
Import/Domestic Price Ratio	0.7	0.9	0.9	0.7	0.8	12.0
Hot-Rolled Sheet						
Import Penetration	0.1	0.1	0.1	0.1	0.1	-24.4
Import/Domestic Price Ratio	0.9	1.0	1.1	0.9	1.1	24.7
Cold-Rolled Sheet						
Import Penetration	0.2	0.1	0.1	0.2	0.2	-16.1
Import/Domestic Price Ratio	0.9	1.0	1.1	0.9	1.1	22.7
Wire Rod						
Import Penetration	0.4	0.3	0.2	0.4	0.4	-13.3
Import/Domestic Price Ratio	0.8	1.0	1.0	1.0	1.1	28.5
Hot-Rolled Bar						
Import Penetration	0.1	0.1	0.1	0.1	0.1	-20.6
Import/Domestic Price Ratio	0.7	1.2	1.0	1.0	1.0	57.8

Source: Compiled from U.S. Department of Commerce, *Current Industrial Reports- Steel Mill Products*, MA33B, Various Years.

considerably for both West Germany and Japan. The impact on iron-ore and coal costs was much smaller than that on labor compensation because both inputs were priced in dollars on the international market. Another factor which contributed to increased Japanese labor costs was falling labor productivity. However, U.S. producer operating costs were only below those of West Germany, although the difference with Japan decreased.

All countries listed in Tables 6.7 to 6.9 increased capacity utilization or maintained high levels. The largest increase was in the U.S., probably due to the more competitive position against countries such as West Germany and Japan, the main countries exporting to the U.S. market. Another factor that contributed to reducing the cost of the U.S. integrated producer was the re-negotiation of coal and iron-ore contracts.

The total cost difference between U.S. operating costs and those of Brazil and South Korea fell from 1986 to 1987 but remained above $100 per ton. The operating-cost disadvantage with South Korea declined from $148 to $121 per ton shipped, while that against Brazil went from $151 to $140. Increasing dollar labor costs in both countries were the main cause of the narrowing of the cost advantage over the U.S. Increased labor productivity was not sufficient to offset the impact of the devaluation on dollar costs per employee. The Brazilians were less affected because of increased capacity utilization; conversely, South Korean finished product yields declined.

110   Steel Production Costs and the Prospects for Slab-Steel Imports

Table 6.7. Cost of Producing Cold-Rolled Sheet at an Efficient Integrated Plant in the United States and Four Selected Countries in 1985 (dollars per metric ton of finished product)

	U.S.	West Germany	Japan	South Korea	Brazil
Iron	74	52	49	53	27
Scrap	20	12			
Coal or Coke	55	53	57	61	75
Other Energy	27	24	17	27	30
Total Raw Materials and Other Energy Costs	175	141	123	141	132
Labor	142	77	70	28	29
Miscellaneous	127	139	124	130	142
Operating Costs	444	357	16	299	303
Depreciation	27	27	32	85	30
Interest	13	17	30	15	88
Total Cost	484	401	378	399	421
Input Prices					
Labor (dollars per man-hour)	22.5	11.9	11.7	2.9	2.9
Iron Ore (dollars per ton)	44.1	28.7	26.7	27.6	13.8
Coal (dollars per ton)	60.6	64.0	65.6	65.1	66.2
Exchange Rate (home currency per dollar)		2.9	240.0	800.0	8500.0
Efficiency Measures					
Man-hour per ton	6.3	6.5	5.9	9.0	9.9
Yield to finished Product (percent)	78.0	80.0	89.0	82.0	80.0
Capacity Utilization	90.0	90.0	90.0	90.0	90.0

Source: Barnett and Crandall, *Up from the Ashes*, 1986.

The data from Tables 6.7 to 6.9 suggest that labor costs are the main advantage of South Korean and Brazilian producers against industrialized countries. The effect of higher labor inputs per ton of steel output is much less than that of lower wages.

As the preceding chapters suggest, the cost differential between the United States and Brazil and South Korea provides incentives for buying steel products from the last two countries. The operating cost advantage is high enough to cover the costs of transporting products from South Korea and Brazil to most of the industrialized countries.

Barnett and Crandall extended their analysis to show the impact of exchange rate differences on 1985 costs. Table 6.10 shows the impact of the different exchange rates, given 90 percent capacity utilization and 1985 input prices and yields. Devaluation of the dollar necessarily produces lower costs for West German and Japanese producers of U. S. dollar-denominated inputs but imposes higher dollar costs of labor and financing.

Table 6.8. Cost of Producing Cold-Rolled Sheet at an Efficient Integrated Plant in the United States and Four Selected Countries in 1986 (dollars per metric ton of finished product)

	U.S.	West Germany	Japan	South Korea	Brazil
Iron	70	48	48	47	23
Scrap	12	11		-5	7
Coal or Coke	52	54	53	64	65
Other Energy	27	21	13	20	23
Total Raw Materials and Other Energy Costs	162	134	114	126	118
Labor	138	98	98	28	27
Miscellaneous	150	178	155	147	150
Operating Costs	449	410	367	301	295
Depreciation	27	38	57	80	30
Interest	13	22	52	15	80
Total Cost	489	470	476	396	405
Input Prices					
Labor (dollars per man-hour)	22.0	15.0	16.7	2.9	2.8
Iron Ore (dollars per ton)	42.0	28.5	24.0	25.0	10.0
Coal (dollars per ton)	60.0	63.0	67.0	69.0	75.0
Exchange Rate (home currency per dollar)		2.3	180.0	850.0	9000.0
Efficiency Measures					
Man-hour per ton	6.3	6.5	5.9	9.0	9.7
Yield to finished product (percent)	80.0	81.5	90.0	86.0	80.0
Capacity Utilization	65.0	70.0	65.0	100.0	90.0

Source: Barnett, *Steel Survival Strategies II*, 1987.

As a result, the cost competitiveness of West German and Japanese producers deteriorates against that of U.S. integrated producers.

As noted, Barnett has estimated how costs differ with the quality of facilities and tried to estimate how the extent of capacity falls into different quality categories. He divides integrated plants into three categories. Each category represents a different degree of new technology adoption. A class 1 producer is one with high product quality and high adoption of critical new technologies. Product quality, adoption of new technology, and operating rates decrease and production costs rise as we move from class 1 to class 3.

No mill in the United States has fully attained class 1 status and, more critically, the distribution of capacity among the three categories differs among product lines. Table 6.11 shows the division in 1987 of U.S. flat-rolled capacity among the three capacity categories. In 1987, 37 percent of integrated crude-steel capacity was class 1, 34 percent was class 2, and 29 percent was class 3.

Table 6.9. Cost of Producing Cold-Rolled Sheet at an Efficient Integrated Plant in the United States and Four Selected Countries in 1987 (dollars per metric ton of finished product)

	U.S.	West Germany	Japan	South Korea	Brazil
Iron Ore	70	57	51	51	30
Scrap	11	11			6
Coal	50	53	55	62	60
Other Energy	24	26	18	20	15
Total Raw Materials and Other Energy Costs	155	147	124	133	111
Labor	143	139	117	44	66
Miscellaneous	147	172	157	147	128
Operating Cost	445	458	398	324	305
Depreciation	28	48	58	77	88
Interest	12	25	53	15	46
Total Cost	485	531	509	416	439
Input Prices:					
Labor (dollars per man-hour)	23.0	22.2	19.0	5.0	5.4
Iron Ore (dollars per ton)	39.7	30.5	25.4	25.4	11.8
Coal (dollars per ton)	51.8	62.7	67.7	67.3	74.6
Exchange Rate (home currency per dollar)		1.8	155.0	800.0	47.1
Efficiency Measures					
Man-hour per ton	6.3	6.3	6.2	8.9	9.3
Yield to finished Product (percent)	83.0	80.0	90.0	82.0	80.0
Capacity Utilization	85.0	79.7	70.0	100.0	100.0

Source: *World Steel Dynamics*, "WSD Cost Monitor," and Barnett, *Steel Survival Strategies III*, 1988.

Of the nine sectors for which data appear in Table 6.11, only electro-galvanizing has the majority of its capacity in class 1. Electro-galvanizing lines built in 1984-87 produced this high level of modernization. The next highest proportion (about 42 percent), and by far the largest absolute amount of class 1 capacity occurs in hot strip mills and hot-dip galvanizing.

Class 1 producers, except for tin mills, have operating rates of at least 80 percent. The main reason for these high operating rates is that their product quality permits them to supply the auto industry under long-term contracts. Tin-mill operating rates have been low because of the competition from substitutes. As noted, operating rates are lower for less modern plants.

The declines differ considerably among product types. Class 2 and class 3 plate capacity utilization differs from utilization of class 1 capacity to a greater extent than is true for other products. Moreover, for plate the difference between class 1 and class 2 utilization rates is much more pronounced than for other product lines shown.

Table 6.10. Impact of Dollar Devaluation Against the Mark and Yen on the 1985 Cost of Producing Cold-Rolled Sheet (dollars per metric ton shipped)

Country	Exchange Rate	Operating Cost	Financial Cost[a]	Total
U.S.		444.3	47.4	491.7
West Germany	3.0	349.5	43.0	392.5
	2.9	357.2	44.1	401.3
	2.5	390.3	50.7	441.0
	2.0	450.9	64.0	514.8
Japan	250.0	308.7	65.1	373.8
	240.0	315.3	67.3	382.6
	200.0	351.7	80.5	432.2
	150.0	425.6	108.1	533.6

Note: [a] Includes taxes costs.
Source: Barnett and Crandall, *Up from the Ashes*, 1986.

Table 6.12 provides comparisons among the production costs, prices, and profit margins of class 1 and 3 producers. Class 1 producers, not only have the lowest cost of production, but also obtain higher prices, which result in a higher profit margin than in class 3. Although the classes of products are the same, the consistency and uniformity of the gage control (thickness consistency), the chemical and the physical properties, and the surface finishing quality, and, therefore, the prices differ. The recovery of steel demand in 1987 allowed both class 1 and 3 producers to earn profits.

The cost effect of dollar depreciation should and did increase imported steel prices from the countries in which the currency appreciated. Table 6.13 shows the effect of the dollar devaluation for the West German and Japanese composite steel prices. Since 1985, both the Japanese and West German prices fell in domestic currency terms but rose in dollar terms. Steel imports from countries under quotas fell below their allowable level (Cantor, 1988). From 1985 to 1986, total steel exports from Japan and West Germany declined by 9 percent for each country (IISI, 1987).

The rise in the imported steel prices contributed to the rise of domestic shipments and plant operating rates, as indicated by the 1987 capacity utilization rates for U.S. integrated producers shown in Table 6.9. As capacity utilization and import prices rose, the "producer price index" for steel mill products increased by 4.7 percent (Cantor, 1988).

Again, a look at market trends gives an alternative view of competitiveness. From 1973 to 1986, import penetration of flat-rolled products (plate, hot-rolled sheet, and cold-rolled sheet), which are predominantly manufactured by integrated producers, increased. The decline of the ratio of import to domestic prices shows that domestic prices are higher than imported prices, which fell for plates and cold-rolled sheets. The increase in the hot-rolled sheet price ratio was influenced by the high demand of the product, which increased the premium for imported steel (Table 6.6).

Table 6.11. U.S. Flat-Rolled Producer 1987 Capacity Assessment

	Class 1	Class 2	Class 3	Total
Capacity				
Coke Ovens	7.3	8.2	9.5	25.0
Blast Furnaces	21.8	15.9	14.5	52.2
Crude Steel	24.5	22.7	19.0	66.2
Slab Casters	4.1	30.8	4.4	39.4
Plate Mills	1.8	2.8	3.4	8.1
Hot-Strip Mills	25.2	16.1	19.3	60.7
Cold Reduction Mills	13.1	10.0	12.7	35.7
Hot-Dip Galvanizing	3.8	3.2	2.3	9.3
Electro-galvanizing	2.2	0.5	0.2	2.8
Tinplates	0.9	2.6	2.3	5.8
Flat-Rolled Products	27.0	19.0	22.8	68.8
Operating Rates[a]				
Coke Ovens	91.0	88.0	80.0	86.0
Blast Furnaces	90.0	85.0	80.0	85.0
Crude Steel	81.0	74.0	60.0	75.0
Slab Casters	88.0	92.0	85.0	89.0
Plate Mills	83.0	75.0	70.0	74.0
Hot-Strip Mills	86.0	80.0	74.0	80.0
Cold Reduction Mills	86.0	81.0	71.0	81.0
Galvanizing Lines	87.0	83.0	74.0	82.0
Tinplates	75.0	70.0	55.0	64.0
Capacity Shares				
Coke Ovens	29.3	32.6	38.0	100.0
Blast Furnaces	41.7	30.4	27.8	100.0
Crude Steel	37.0	34.2	28.8	100.0
Slab Casters	10.4	78.3	11.3	100.0
Plate Mills	22.5	34.8	42.7	100.0
Hot-Strip Mills	41.6	26.6	31.8	100.0
Cold Reduction Mills	36.5	27.9	35.5	100.0
Hot-Dip Galvanizing	41.2	34.3	24.5	100.0
Electro-galvanizing	77.4	16.1	6.5	100.0
Tinplates	15.6	45.3	39.1	100.0
Flat-Rolled Capacity	39.3	27.6	33.1	100.0

Note: [a] First half-year tonnages have been annualized.
Source: Barnett, *Steel Survival Strategies II*, 1987.

THE COMPETITIVENESS OF PURCHASING SLAB OR DRI BY U.S. FLAT-ROLLED CARBON STEEL PRODUCERS

This section provides a more detailed discussion of slab import economics. Table 3.22 reports carbon steel slab import prices, and the data help to identify low-cost suppliers. In most years shown, Belgium, Brazil, South Korea, United Kingdom, and South Africa provided the lowest

Table 6.12. 1987 Costs of Production, Prices, and Profit Margin for U.S. Class 1 and 3 Integrated Producers (dollars per metric ton shipped)

	Production Cost		Price		Margin	
	Class 3	Class 1	Class 3	Class 1	Class 3	Class 1
Hot-Rolled Sheet						
Commercial Quality	320	287	347	375	28	88
Drawing Quality	331	298	358	386	28	88
Extra Drawing Quality	342	309	369	397	28	88
Cold-Rolled Sheet						
Commercial Quality	447	397	480	507	39	110
Drawing Quality	463	413	496	524	39	110
Extra Drawing Quality	480	430	513	540	39	110
Galvanized Hot-Dipped						
Commercial Quality	551	491	584	628	33	138
Drawing Quality	573	513	617	662	44	149
Extra Drawing Quality	595	535	650	695	55	160
Electro-galvanizing Sheet	606	540	662	673	55	132
Tin Plate	650	606	695	717	44	110

Source: Barnett, *Steel Survival Strategies II*, 1987.

priced supplies. The highest prices for slab were for those from Sweden, West Germany, France, and the Netherlands. In 1988, the price of slab exports to the United States from the Netherlands fell such an extent that the Netherlands became a low-cost supplier, and Australia began to supply the U.S. market as a low-cost supplier. Brazil is the only significant supplier of slab with consistently low prices. The potential for Venezuela and Mexico to become low price, major slab suppliers is indicated by the growth of their exports to the United States at low prices. As discussed more fully in the rest of this chapter, Latin America could become a major supplier of slab to the U.S. market, due to low shipping ($10 to $20 per ton, depending on the exporting country) and operating costs.

An Analysis of the Competitiveness of U.S. Slab Production

To appraise the international cost competitiveness of the United States and other countries in producing slabs, the accounting cost of producing slab was estimated using the procedure outlined in Appendix D. Given the

Table 6.13. Home Country Composite Steel Prices of the United States, West Germany, and Japan in Dollar and Home Countries' Currency--1980 to 1987

	U.S. dollar (per ton)	Japan (dollar per ton)	Japan	West Germany	West Germany
1980	491	429	96,726	481	875
1981	555	442	97,279	389	880
1982	554	391	97,040	414	1,004
1983	514	418	99,256	389	993
1984	525	415	98,527	353	1,005
1985	500	407	96,368	353	1,040
1986	487	548	92,019	461	1,001
1987[a]	510	591	86,770	492	906

Note: [a] 1987 prices refers to the third quarter. All figures are composite and assume a U.S. major mill 1977 carbon steel shipment mix. Home prices are f.o.b.

Source: *World Steel Dynamics*, "Steel Strategist #14," 1987.

importance discussed above of transportation costs, the analysis is conducted on a delivered price basis. In particular, the analysis discusses the competitive position of U.S. mills in three locations serving three markets.

The mills are on the East Coast, the West Coast, and an inland location such as Chicago. While only the ability of coastal mills to compete in nearby markets is considered, the midwestern mill's competitive position on both coasts is examined. This requires calculating the delivered cost of slabs to these plants, the delivered cost of flat-rolled products produced from different sources of slab in the markets, and the cost of imported hot-rolled steel delivered to these markets.

Tables 6.14 and 6.15 show the 1985 and 1988 operating costs and total cost estimates for producing slab at efficient integrated producers in the foreign countries listed in Table 6.9 and for U.S. integrated producers at the locations considered. The years 1985 and 1988 were selected because of the extremes in the value of the U.S. for West German and Japanese producers compared to other strong currencies and in U.S. and international demand for steel products. The inputs and equipment efficiencies considered for each country refer to efficient integrated plants that may be different from those viewed in Tables 6.7 to 6.9.

In this analysis, the only difference in U.S. production costs inland compared to those at a coastal location is in iron ore. Coastal integrated producers have a lower cost than those located inland because of lower transportation costs.

In 1985, the U.S. inland integrated producer had the highest operating cost and total cost of all producers shown, and the U.S. coastal integrated producer was the only one competitive with the West Germans. Both U.S. and West German domestic currency costs fell from 1985 to 1988. Even

Table 6.14. Cost of Producing Slab at Efficient Integrated Producers for U.S. Inland and Coastal Regions and Four Selected Countries in 1985 (dollars per metric ton)

	U.S. Inland	U.S. Coastal	West Germany	Japan	South Korea	Brazil
Iron Ore	43	34	49	36	43	23
Scrap	26	26	30	7	6	10
Coal	47	47	45	52	58	54
Other Energy	1	1	-1	-7	-1	1
Total Raw Materials and Other Energy Costs	117	108	122	88	106	88
Labor	78	78	42	36	13	13
Miscellaneous	89	89	114	79	96	95
Operating Cost	284	275	278	204	215	196
Depreciation	26	26	26	41	77	24
Interest	13	13	17	37	14	22
Taxes	8	8	1	8	1	1
Total Cost	331	322	322	290	308	244
Input Prices						
Labor (dollars per man-hour)	22.5	22.5	11.9	11.7	2.9	2.9
Iron Ore (dollars per metric ton)	33.0	26.2	28.7	26.7	27.6	13.8
Coal (dollars per metric ton)	60.6	60.6	64.0	65.6	65.1	66.2
Exchange Rate (home currency per dollar)			2.9	240	800	8,500
Efficiency Measures						
Continuous Casting	80	80	80	100	60	60
Man-Hours per Ton	3.5	3.5	3.5	3.1	4.8	4.6

Source: Calculated using the cost estimation procedure of Appendix D.

with mark appreciation, German costs in U.S. dollars were lower in 1988 than in 1985, but U.S. costs fell even more. An increase in continuous casting, higher labor productivity, and reduction of the coke rate were major influences on German costs.

Japanese production costs rose from $290 to $302 per ton. Japan had attained 100 percent continuous casting in 1985, and its coke rate rose as oil injection was eliminated. As a result, Japanese U.S. dollar costs rose as the yen appreciated against the dollar.

Efficiency improvements by U.S. integrated firms produced lower production costs for both coastal and inland mills. From 1985 to 1988, U.S. producers increased continuous casting production from 80 percent to 90 percent, and labor use improved from 3.47 to 2.87 man-hours per ton. Another contribution was made by reduced coke and iron-ore costs. In 1988, U.S. slab-production operating cost throughout the United States ranged from $220 to $265 per ton (*World Steel Dynamics*, 1988).

The differential in total costs between the U.S. integrated plants and the two developing countries diminished but did not disappear. The cost

Table 6.15. Cost of Producing Slab at Efficient Integrated Producers for U.S. Inland and Coastal Regions and Four Selected Countries in 1988 (dollars per metric ton)

	U.S. Inland	U.S. Coastal	West Germany	Japan	South Korea	Brazil
Iron Ore	50	41	46	37	40	23
Scrap	31	31	26	10	22	11
Coal	32	32	40	47	46	44
Other Energy	2	2	-1	-6	-1	1
Total Raw Materials and Other Energy Costs	115	106	111	89	107	79
Labor	60	60	62	44	15	17
Miscellaneous	74	74	96	89	84	92
Operating Cost	248	240	269	223	206	187
Depreciation	25	25	30	40	55	24
Interest	10	10	15	33	11	15
Taxes	3	3	1	7	1	1
Total Cost	287	278	315	302	273	228
Input Prices						
Labor (dollars per metric ton)	26.5	26.5	24.3	19.9	5.0	4.4
Iron Ore (dollars per metric ton)	35.3	29.0	29.8	24.3	25.9	13.8
Coal (dollars per metric ton)	55.1	55.1	62.8	61.7	61.7	61.2
Exchange Rate (home currency per dollar)			1.7	135	800	110
Efficiency Measures						
Continuous Casting	90	90	90	100	80	70
Man-Hour per Ton	2	2	3	2	3	4

Source: Calculated using the cost estimation procedure of Appendix D.

advantage of a South Korean producer declined from $23 to $14 per ton compared to a U.S. integrated producer located inland and from $14 to $5 per ton compared to a coastal producer. The cost edge of a Brazilian producer declined from $87 to $59 per ton compared to integrated producer located inland and from $78 to $50 per ton compared to the coastal. In 1988, coastal total production cost was almost comparable to that of South Korea. This was not the case with a Brazilian producer because of cheap iron ore. The reduction of the operating-cost disadvantage for U.S. integrated producers relative to those in the two developing countries was smaller.

Brazil and South Korea have a cost advantage in the production of slab because labor costs are significantly lower in those countries than in the United States, and in the case of Brazil, because of cheap iron ore, too. Moreover, changes in exchange rates do not significantly affect the cost competitiveness of these countries because their currencies are kept closely tied to the U.S. dollar.

Competitiveness in Producing Flat-Rolled Products

This section compares the cost of producing a fixed product mix of flat-rolled products consisting of 30 percent hot-rolled sheet, 30 percent cold-rolled sheet, and 40 percent galvanized sheet. The flat-rolled producer produces 4 million tons of finished flat-rolled products. As noted in an earlier section of this chapter, relating production costs estimates to the entire product mix eliminates cost allocation problems.

To appraise locational influences, the following situation was evaluated:

--two domestic slab suppliers, one located inland and the other located in the East (coastal), since no efficient slab producer exists in the West;
--two foreign slab suppliers, South Korea and Brazil; and
--three U.S. slab purchasers with efficient and inefficient finishing-end located inland (Chicago area), in the West, and in the East.

Table 6.16 shows the total cost for transporting slabs from the supplier to the plant in the different regions considered. Table 6.17 shows the delivered price from each slab producer considered in Tables 6.14 and 6.15 to each of the three U.S. locations treated. Transportation costs from Table 6.16 are added to the operating and total costs in Tables 6.14 and 6.15.

Slab prices when slab producers recover operating costs were estimated because in a year of low steel-product demand (such as 1985) slab prices fall to levels than can be below operating costs as measured here. (Some of the operating costs may be fixed costs to producers vertically integrated into iron-ore and coal production.) Slab prices when slab producers recover total cost were estimated because, as demand for steel products increases (as in 1988), slab producers tend to cover total costs.

The delivered price for Brazilian slab is the lowest of all regions. The second best alternative varies with the region supplied and with the extent to which slab producers recover operating and total costs. In the West, South Korean slab is the second-best option and, in 1985, was the second-best option for both inland and Eastern slab purchasers. In 1988, the domestic slab producer closer to the region supplied was the second-best option. These basic relationships are a primary determinant of the cost and profitability comparisons presented in the rest of this chapter.

Tables 6.18 and 6.19 show the estimated total composite production cost and profit margin at a non-integrated slab purchaser with an efficient and an inefficient finishing-end. As expected, the non-integrated slab purchaser with an efficient finishing-end has a lower production cost than the one with an inefficient finishing-end. For the purpose of making these calculations, materials yield and maintenance costs are assumed to be the prime source of cost differences on both the hot metal and finishing side. Hot metal costs also depend on the use of continuous casting. My sources indicated the range of costs and yields prevailing. My efficient plants had maintenance costs at the low end of the estimated cost range and materials yields at the high end. The inefficient plants then had maintenance costs at the high end of the range and yields at the low end. On the hot metal side an efficient

Table 6.16. Transportation Cost from Exporting Regions and Domestic Producers to Selected Regions in the United States in 1988-89 (dollars per metric ton)

	Delivered Region		
	West	East	Inland[b]
From Domestic Producer to Delivered Region			
Inland	45.0	30.0	10.0
Coastal (East)	45.0	10.0	30.0
From Selected Countries to U.S. Port			
Slabs			
Brazil	23.0	17.5	26.5
South Korea	17.5	23.0	32.5
Finished Products			
Brazil	46.0	34.0	40.0
South Korea	34.0	39.0	46.0
From U.S. Port to Delivered Region			
Same for All Countries and Products	20.0	17.5	14.5

Note: [a] The slab shipping cost are based on contracting a vessel of large tonnage and insurance.
[b] From foreign producers by shipping on the Mississippi River.
Source: Discussions with trading companies and steel producers.

plant also continuously cast all its output but only 45 percent of output at inefficient plants was continuously cast.

In 1985, with slab prices equal to operating costs, the efficient non-integrated mills at all locations purchasing slab from Brazil were the only ones to show a positive profit margin. In 1988 with slab prices equal to total costs, all non-integrated producers in all regions showed the highest profit margin by purchasing slab from Brazil.

In 1988, an efficient non-integrated producer at all locations was profitable when purchasing slab from South Korea. When employing domestic suppliers, efficient purchasers located inland and in the East were profitable. In the same year, inefficient non-integrated producers were profitable only in the East and inland when purchasing slab from the closest domestic supplier.

The Western slab purchaser is only profitable when importing slabs. Therefore, the Western efficient non-integrated mill may only purchase domestic slab supplies in order to maintain full utilization of its plant although reducing its profit margin.

The cost competitiveness of imported slab supplies from Brazil and South Korea arises primarily from the lower costs of production. On the West Coast, foreign suppliers also have a transportation cost advantage.

To evaluate the impact of imported slab and of producing flat-rolled products at an efficient and an inefficient non-integrated plant, the cost of producing the same product mix at an integrated plant by region, type of plant, and iron-making technology used was estimated. Tables 6.20 and

Table 6.17. Estimated Slab Price Delivered at the Plant Location in 1985 and 1988 (dollars per metric ton)[a]

	Plant Region in United States		
	West	East	Inland
When Slab Producer Recovers Operating Cost			
1985			
Brazil	239.0	231.0	237.0
South Korea	252.5	255.5	262.0
Domestic Inland	329.0	314.0	294.0
Domestic Coastal	320.0	285.0	305.0
1988			
Brazil	230.0	222.0	228.0
South Korea	243.0	246.5	253.0
Domestic Inland	293.0	278.0	258.0
Domestic Coastal	285.0	250.0	270.0
When Slab Producer Recovers Total Cost			
1985			
Brazil	287.0	279.0	285.0
South Korea	345.5	348.5	355.0
Domestic Inland	376.0	361.0	341.0
Domestic Coastal	367.0	332.0	352.0
1988			
Brazil	271.0	263.0	269.0
South Korea	310.5	313.5	320.0
Domestic Inland	332.0	317.0	297.0
Domestic Coastal	323.0	288.0	308.0

Note: [a] This price is estimated from adding operating and total production cost to transportation cost from slab producer to purchasing plant
Source: Compiled from Tables 6.14, 6.15, and 6.16.

6.21 show the total composite production cost and profit margin for all these alternative production patterns.

In 1985, non-integrated producers located inland and in the East purchasing slab from Brazil were more profitable than were the most efficient integrated producer using conventional iron-making technology. In fact, use of any type of conventional technology produced losses.

In 1988, integrated producers with efficient hot-end operations were profitable whatever the efficiency of the finishing-end. Only the efficient non-integrated producer purchasing slab from Brazil enjoyed a comparable or better result than integrated producers using conventional iron-making technology located inland and in the East.

The tables also show the effect of new iron-making technology, namely Corex, on costs and profits. If Corex costs are as low as its advocates claim, the profitability of integrated operations is increased to an extent to which importing slab becomes less attractive. Such technology, as noted in Chapter 2, could allow non-integrated producers to integrate backward into

Table 6.18. Estimated Total Composite Production Cost at a U.S. Efficient and Inefficient Non-Integrated Producer Purchasing Slab (dollars per metric ton)[a] [b]

	Efficient Non-Integrated			Inefficient Non-Integrated		
	West	East	Inland	West	East	Inland
When Slab Producer Recovers Operating Cost						
1985						
Brazil	472.5	463.6	470.3	493.6	484.0	491.2
South Korea	487.5	490.9	498.1	509.8	513.4	521.2
Domestic Inland	572.6	555.9	533.7	601.7	583.7	559.7
Domestic Coastal	562.6	523.7	545.9	590.9	548.8	572.9
1988						
Brazil	465.9	457.0	463.7	485.9	476.4	483.5
South Korea	481.0	484.3	491.5	502.2	505.8	513.6
Domestic Inland	536.0	519.3	497.1	561.6	543.6	519.6
Domestic Coastal	527.1	488.2	510.4	552.0	510.0	534.0
When Slab Producer Recovers Total Cost						
1985						
Brazil	525.9	517.0	523.7	551.2	541.6	548.8
South Korea	591.0	594.3	601.5	621.5	625.1	632.9
Domestic Inland	624.9	608.2	585.9	658.1	640.1	616.1
Domestic Coastal	614.9	575.9	598.2	647.3	605.3	629.3
1988						
Brazil	511.5	502.6	509.3	535.2	525.6	532.8
South Korea	555.5	558.8	566.0	582.6	586.2	594.0
Domestic Inland	579.4	562.7	540.4	608.4	590.4	566.4
Domestic Coastal	569.4	530.4	552.7	597.6	555.6	579.6

Note: [a] Composite cost estimates throughout this chapter are based on a plant product mix of 30 percent hot-rolled sheet, 30 percent cold rolled sheet, and 40 percent galvanized sheet.
[b] Both non-integrated producers employ non-unionized labor in 1985 US$ 17.50 per man-hour) and in 1988 (US$ 19.00 per man-hour).
Source: Calculated using the cost estimation procedure of Appendix D.

iron-making. Corex would also provide another way for efficient finishing-end facilities to survive.

The profitability of investing in new iron-making technology, closing hot-end facilities, and purchasing imported slabs for the efficient finishing-end facilities is discussed in the next section.

THE PROFITABILITY OF AN INTEGRATED PRODUCER AND A NON-INTEGRATED SLAB PURCHASER

This part of the chapter examines the profitability of different investment options available to existing flat-rolled producers and new entrants considering two different scenarios for flat-rolled products. Table 6.22

Table 6.19 Profit Margin at a U.S. Efficient and Inefficient Non-Integrated Producer Purchasing Slab (dollars per metric ton)[a]

	Efficient Non-Integrated			Inefficient Non-Integrated		
	West	East	Inland	West	East	Inland
When Slab Producer Recovers Operating Cost						
1985						
Brazil	9.5	18.4	11.7	-11.6	-2.00	-9.2
South Korea	-5.5	-8.9	-16.1	-27.8	-31.4	-39.2
Domestic Inland	-90.6	-73.9	-51.7	-119.7	-101.7	-77.7
Domestic Coastal	-80.6	-41.7	-63.9	-108.9	-66.8	-90.9
1988						
Brazil	101.1	103.3	108.3	81.1	90.6	83.5
South Korea	86.1	82.7	75.5	64.9	61.3	53.4
Domestic Inland	31.0	47.7	69.9	5.4	23.4	47.4
Domestic Coastal	39.9	78.8	56.6	15.0	57.0	33.0
When Slab Producer Recovers Total Cost						
1985						
Brazil	-43.9	-35.0	-41.7	-69.2	-59.6	-66.8
South Korea	-109.0	-112.3	-119.5	-139.5	-143.1	-150.9
Domestic Inland	-142.9	-126.2	-103.9	-176.1	-158.1	-134.1
Domestic Coastal	-132.9	-93.9	-116.2	-165.3	-123.3	-147.3
1988						
Brazil	55.5	64.4	57.7	31.8	41.4	34.2
South Korea	11.6	8.2	1.0	-15.6	-19.2	-27.0
Domestic Inland	-12.4	4.3	26.6	-41.4	-23.4	0.6
Domestic Coastal	-2.4	36.6	14.3	-30.6	11.4	-12.6

Note: [a] Interviews with steel producers and traders indicate that composite selling prices are $482 per metric ton in 1985 and $567 per metric ton in 1988. These prices relate to actual market prices of the product mix used in the total production cost estimates.

Source: Compiled from Table 6.18 and composite prices collected from steel traders and producers.

summarizes the estimated investment cost, life, construction time, and producer expectations associated with these investment alternatives.

The conservative producer perceives the future as having an equal number of bad years (epitomized by 1985) and good years (epitomized by 1988). An optimist producer perceives the future as consisting entirely of good years.

All investment options involve the same product mix and finished flat-rolled capacity considered above. Therefore, the expected profit margin for the conservative producer is the average of the 1985 and 1988 profit margins, and the expected profit margin for the optimist producer is the 1988 profit margin. In addition, two discount rates, 6 percent and 8 percent, are used in calculating the net present value estimates. Calculations are made only for options that at least cover operating costs and thus can repay at least some investment costs.

Table 6.20. Estimated Total Composite Production Costs at a U.S. Integrated Producer by Region, Iron-Making Technology Used, and Type of Plant (dollars per metric ton)[a]

	1985	1988
Conventional Iron-Making Technology		
U.S. Coastal Integrated Producer		
Efficient Hot-End and Finishing-End	493.5	509.4
Efficient Hot-End and Inefficient Finishing-End	518.4	534.0
Inefficient Hot-End and Efficient Finishing-End	594.4	611.0
Inefficient Hot-End and Finishing-End	620.6	637.1
U.S. Inland Integrated Producer		
Efficient Hot-End and Finishing-End	500.3	515.2
Efficient Hot-End and Inefficient Finishing-End	525.7	540.3
Inefficient Hot-End and Efficient Finishing-End	602.1	617.7
Inefficient Hot-End and Finishing-End	628.9	644.2
New Iron-Making Technology (Corex)[b]		
U.S. Coastal Integrated Producer		
Efficient Finishing-End	460.9	476.6
Inefficient Finishing-End	479.2	494.9
U.S. Inland Integrated Producer		
Efficient Finishing-End	469.7	489.8
Inefficient Finishing-End	488.7	513.8

Note: [a] Composite cost estimates throughout this chapter are based on a plant product mix of 30 percent hot-rolled sheet, 30 percent cold rolled sheet, and 40 percent galvanized sheet.

[b] The cost of Corex hot metal was estimated for the coastal plant in 1985 and 1988 at $85 and $90 per metric ton and for the inland plant at $95 and $105 per metric ton, respectively.

Source: Calculated using the cost estimate procedure of Appendix D.

Tables 6.23 and 6.24 measure the profitability of new entrants into the flat-rolled market by (1) investing in a green-field (new) plant using different iron-making technologies, (2) investing in a green-field non-integrated plant, or (3) buying an existing plant with efficient or inefficient finishing-end. The new entrant would not consider building a green-field integrated plant by using either conventional or new iron-making technologies even if it behaves optimistically.

Investment in a green-field non-integrated plant would be expected to be profitable only if the firm behaves as an optimist, accepts a 6 percent discount rate, and secures slab supplies from Brazil. Although this investment is profitable for all regions, the most probable location would be in the East for imported Brazilian slabs. This alternative is viable mainly because it reduces the total investment cost by 67 percent.

Table 6.21. Estimated Profit Margin at a U.S. Integrated Producer by Region, Iron-Making Technology Used, and Type of Plant (dollars per metric ton)[a]

	1985	1988
Conventional Iron-Making Technology		
U.S. Coastal Integrated Producer		
Efficient Hot-End and Finishing-End	-11.5	57.6
Efficient Hot-End and Inefficient Finishing-End	-36.4	33.0
Inefficient Hot-End and Efficient Finishing-End	-112.4	-44.0
Inefficient Hot-End and Finishing-End	-138.6	-70.1
U.S. Inland Integrated Producer		
Efficient Hot-End and Finishing-End	-18.3	51.8
Efficient Hot-End and Inefficient Finishing-End	-43.7	26.7
Inefficient Hot-End and Efficient Finishing-End	-120.1	-50.7
Inefficient Hot-End and Finishing-End	-146.9	-77.2
New Iron-Making Technology (Corex)		
Coastal U.S. Integrated Producer		
Efficient Finishing-End	21.1	90.4
Inefficient Finishing-End	2.8	72.1
Inland U.S. Integrated Producer		
Efficient Finishing-End	-11.3	77.2
Inefficient Finishing-End	-6.7	53.2

Note: [a] Interviews with steel producers and traders indicate that composite selling prices were $482 per metric ton in 1985 and $567 per metric ton in 1988. These prices relate to actual market prices of the product mix used in the total production cost estimates.

Source: Compiled from Table 6.20 and composite prices collected from steel traders and producers.

All discussions here of purchasing plants assume the price will be similar to those observed on sales made in the 1980s. Economic theory predicts that the selling price of a facility reflects its current profit prospects. Therefore, plants incorporating facilities purchased at the prices that have prevailed on steel-plant sales during the 1980s would be expected to be more profitable than totally new plants. For example, this is true for a non-integrated plant with finishing facilities purchased from a closed plant and refurbished, as is the case of Tuscaloosa.

Success of a green-field integrated plant might be possible only if new iron-making technology and new flat-rolled technology, such as in thin-slab casting, are developed and have sufficiently low costs. This alternative was not analyzed here because the necessary data were not available. However, Barnett estimates that profit margins for hot-rolled sheets would increase by $75 per ton for a class 3 producer and $20 per ton for a class 1 producer. He expects increases in profit margins in cold-rolled sheets to be $105 per ton for a class 3 producer and $40 per ton of class 1 producer (Barnett, 1987).

Table 6.22. Estimates of Investment Options of a U.S. Flat-Rolled Producer and Its Behavior

Investment[a]	Life of Project (years)	Construction Time (years)	Initial Investment (billion of dollars)
Greenfield Integrated Plant with Conventional Iron-Making Technology[b]	24	4	5,815
Greenfield Integrated Plant with New Iron-Making Technology[b]	24	4	3,119
Greenfield Non-Integrated Plant Purchasing Slab	24	2	2,319
Conventional Iron-Making[b]	22	0	1,119
Conventional Iron-Making and Continuous Casting[b]	22	0	1,455
New Iron-Making and Continuous Casting[b]	22	0	800
Closing Hot-End Facilities	22	0	455
Purchasing an Existing Plant with Efficient Finishing-End	22	0	400
Purchasing an Existing Plant with Inefficient Finishing-End	22	0	100

Flat-Rolled Producer Behavior	Probability (percent)	
Scenario	Bad Year 1985	Good Year 1988
Conservative	50	50
Optimist	0	100

Note: [a] Composite estimates throughout this chapter are based on a plant product mix of 30 percent hot-rolled sheet, 30 percent of cold-rolled sheet, and 40 percent galvanized sheet.

[b] All these alternatives considered that blast furnaces and Corex are relined every six years at a cost of $60 per metric ton remaining inactive for one year. During this year of inactivity finishing facilities consumes the cheapest slab supplies available.

Source: Discussions with trading companies and steel producers and *World Steel Dynamics,* "Steel Survival Strategies III," 1988.

Two types of plants, one with efficient finishing-end facilities, and another with inefficient facilities, theoretically are available to a new entrant buying an existing plant. With either type of demand expectation and finishing-mill efficiency, importing Brazilian slab is the cheapest supply source in each of the three U.S. regions considered.

An optimistic firm in the West finds its second-best option is to buy a plant with an efficient finishing-end and to purchase slabs from South Korea. At the other two U.S. locations, an optimistic entrant would find

Table 6.23. Net Present Value of a Green-Field Plant Producing Flat-Rolled Products by Type of Plant for Selected U.S. Regions Producer Behavior at 6 Percent and 8 Percent Discount Rates[a] [b]

U.S. Green-Field Integrated Plant	Conservative	Optimist
Using Conventional Iron-Making Technology		
U.S. Coastal Integrated Producer		
6 Percent Rate	-5,097	-3,881
8 Percent Rate	-5,242	-4,276
U.S. Inland Integrated Producer		
6 Percent Rate	-5,310	-4,078
8 Percent Rate	-5,411	-4,432
Using New Iron-Making Technology (Corex)		
U.S. Coastal Integrated Producer		
6 Percent Rate	-1,328	-191
8 Percent Rate	-1,692	-786
U.S. Inland Integrated Producer		
6 Percent Rate	-1,608	-543
8 Percent Rate	1,917	-1,068

U.S. Green-Field Non-Integrated Plant	Conservative			Optimist		
Plant Region	West	East	Inland	West	East	Inland
6 Percent Rate						
Brazil	-927	-545	-833	59	440	154
South Korea	-2,190			-1,824	-1,967	
Domestic Inland					-2,134	-1,180
Domestic Coast					-751	-1,705
8 Percent Discount Rate						
Brazil	-1,183	-871	-1,106	-379	-67	-301
South Korea	-2,214			-1,915	-2,032	
Domestic Inland					-2,168	-1,390
Domestic Coast					-1,040	-1,818

Note: [a] Composite estimates throughout this chapter are based on a plant product mix of 30 percent hot-rolled sheet, 30 percent cold rolled sheet, and 40 percent galvanized sheet.

[b] Net present values in the conservative case use the average of profits in the case in which producers only recover operating costs as indicated by 1985 profit margins and in the case in which producers recover total costs as indicated by the 1988 profit margins. The present values in the optimistic case are based on 1988 profit margins. Discussions with steel constultants, producers, and traders indicate that this assumption produces the best estimate of slab delivered price in 1985.

Source: Results from net present value analysis.

buying (at the prices prevailing in the 1980s) an efficient finishing mill and buying slab produced in the region in which the finishing mill is located the second best option. Buying a plant with an inefficient finishing-end

Table 6.24. Net Present Value of Buying an Existing Integrated Plant with Efficient and Inefficient Finishing-End for Selected U.S. Regions and Producer Behavior at 6 Percent and 8 Percent Discount Rate[a] [b]

Efficient Finishing-End Plant Region	Conservative			Optimist		
	West	East	Inland	West	East	Inland
6 Percent Rate						
Brazil	1,164	1,593	1,270	2,272	2,700	2,379
South Korea	-255			156	-5	-111
Domestic Inland					-192	879
Domestic Coast					1,361	290
8 Percent Rate						
Brazil	925	1,288	1,015	1,863	2,226	1,958
South Korea	-277			71	-65	-156
Domestic Inland					-244	684
Domestic Coast					1,092	185

Inefficient Finishing-End Plant Region	Conservative			Optimist		
	West	East	Inland	West	East	Inland
6 Percent Rate						
Brazil	387	850	503	1,433	1,896	1,549
Domestic Inland						-71
Domestic Coast					450	
8 Percent Rate						
Brazil	313	705	631	1,199	1,590	1,517
Domestic Inland						-76
Domestic Coast					366	

Note: [a] Composite estimates throughout this chapter are based on a plant product mix of 30 percent hot-rolled sheet, 30 percent cold rolled sheet, and 40 percent galvanized sheet.

[b] Net present values in the conservative case use the average of profits in the case in which producers only recover operating costs as indicated by 1985 profit margins and in the case in which producers recover total costs as indicated by the 1988 profit margins. The present values in the optimistic case are based on 1988 profit margins. Discussions with steel constultants, producers, and traders indicate that this assumption produces the best estimate of slab delivered price in 1985.

Source: Results from net present value analysis.

supplied by a local domestic slab supplier would appear profitable only in the East.

In sum, only an optimistic producer using Brazilian slab would expect an green-field non-integrated plant to be profitable. Purchasing (again at the prices of the 1980s) an efficient existing finishing mill and using Brazilian slab would be expected to be the most profitable alternative in any region of the U.S. under either demand expectation. Only an optimist would expect

Steel Production Costs and the Prospects for Slab-Steel Imports

some other patterns of purchasing existing plants and securing slab to be profitable.

Tables 6.25 and 6.26 estimate the profitability and survival of existing integrated plants with (1) closing hot-end facilities, (2) substituting hot-end facilities with conventional iron-making technology at an integrated steel plant with present 100 percent continuous casting adoption or with present 45 percent continuous casting adoption, or (3) substituting hot-end facilities with new iron-making technology in a plant with 45 percent present continuous casting adoption.

A conservative producer only expects profitability by closing hot-end facilities and being reorganized as a non-integrated producer purchasing slab from Brazil. An optimistic producer finds that the use of Brazilian slab is his lowest cost alternative but expects to be profitable substituting Korean slab on the West Coast and locally produced slab on the East Coast and inland. These results indicate the attractiveness to Lone Star of closing its hot-end facilities.

The options appraised in Table 6.26 involve only replacing part of an existing plant and, therefore, are inherently less costly that an entirely new plant. Such replacements might be and, in some cases, appear to be profitable even when a completely new plant is uneconomic. An existing integrated plant with 100 percent continuous casting would replace its iron-making facilities in all situations considered, except for a plant located inland considering a discount rate of 8 percent and with a producer behaving conservatively. The plant with an inefficient hot-end will only undertake the investment of substituting its iron-making facilities and increasing continuous casting if the producer is an optimist at both locations. If the producer is conservative, its survival depends on the availability of new iron-making technologies.

THE COMPETITIVENESS OF IMPORTING SLAB

This section examines the international competitiveness of U.S. integrated and non-integrated slab purchasers in the flat-rolled market. To appraise the international competitiveness of U.S. flat-rolled producers, imports from Brazil and South Korea of a product mix identical to that considered above is examined. These two exporting countries were used because they appear to be the most competitive producers of flat-rolled products.

Table 6.27 shows the estimated total cost at three delivered regions, in 1988, of U.S. flat-rolled producers and the two exporting countries, where the U.S. flat-rolled delivered cost is at the plant site, and where the two exporting countries' cost is that unloaded at the port. The cost from these two countries is lower than U.S. flat-rolled producers in all regions, except for Brazilian imports that compete with U.S. Western non-integrated slab purchasers.

Table 6.25. Net Present Value of Closing the Hot-End at a Plant with Efficient and Inefficient Finishing-End for Selected U.S. Regions and Producer Behavior at 6 Percent and 8 Percent Discount Rate[a] [b]

Efficient Finishing-End Plant Region	Conservative			Optimist		
	West	East	Inland	West	East	Inland
6 Percent Rate						
Brazil	1,119	1,548	1,215	2,227	2,655	2,324
South Korea	-300			111	-50	-156
Domestic Inland					-237	834
Domestic Coast					1,316	245
8 Percent Rate						
Brazil	880	1,243	960	1,818	2,181	1899
South Korea	-322			26	-110	-201
Domestic Inland					-269	639
Domestic Coast					1,047	140

Inefficient Finishing-End Plant Region	Conservative			Optimist		
	West	East	Inland	West	East	Inland
6 Percent Rate						
Brazil	42	505	148	1,018	1,551	1,194
Domestic Inland						-416
Domestic Coast				105		
8 Percent Rate						
Brazil	-32	360	56	854	1,245	942
Domestic Inland						-421
Domestic Coast				21		

Note: [a] Composite estimates throughout this chapter are based on a plant product mix of 30 percent hot-rolled sheet, 30 percent cold rolled sheet, and 40 percent galvanized sheet.

[b] Net present values in the conservative case use the average of profits in the case in which producers only recover operating costs as indicated by 1985 profit margins and in the case in which producers recover total costs as indicated by the 1988 profit margins. The present values in the optimistic case are based on 1988 profit margins. Discussions with steel constultants, producers, and traders indicate that this assumption produces the best estimate of slab delivered price in 1985.

Source: Results from net present value analysis.

The margins between the two exporting countries and U.S. flat-producers vary according to the region considered. Inland, the margin is in the $7 to $8 per ton range compared to both efficient non-integrated and integrated. In the East, the margin is in the $3 to $5 per ton range compared to U.S. efficient non-integrated firms and $16 to $21 per ton compared to U.S. integrated producers. In the West, the margin is the -$2 to $16 per ton range compared to U.S. efficient non-integrated companies, and $41 to $61

Table 6.26. Net Present Value of Replacing Hot-End Facilities at an Existing Integrated Plant with an Efficient Finishing-End for Selected U.S. Regions and Producer Behavior at 6 Percent and 8 Percent Discount Rate[a] [b]

	Conservative	Optimist
Replacing with Coke Oven and Blast Furnace		
U.S. Coastal Integrated Producer		
6 Percent Rate	330	2,023
8 Percent Rate	128	1,525
U.S. Inland Integrated Producer		
6 Percent Rate	67	1,517
8 Percent Rate	-85	1,118
Replacing with Coke Oven, Blast Furnace and Continuous Casting[c]		
U.S. Coastal Integrated Producer		
6 Percent Rate	-51	1,642
8 Percent Rate	-253	1,144
U.S. Inland Integrated Producer		
6 Percent Rate	-314	1,136
8 Percent Rate	-466	737
Replacing with New Iron Making (Corex) and Continuous Casting[c]		
U.S. Coastal Integrated Producer		
6 Percent Rate	991	2,128
8 Percent Rate	627	1,533
U.S. Inland Integrated Producer		
6 Percent Rate	711	1,776
8 Percent Rate	402	1,251

Note: [a] Composite estimates throughout this chapter are based on a plant product mix of 30 percent hot-rolled sheet, 30 percent cold rolled sheet, and 40 percent galvanized sheet.

[b] Net present values in the conservative case use the average of profits in the case in which producers only recover operating costs as indicated by 1985 profit margins and in the case in which producers recover total costs as indicated by the 1988 profit margins. The present values in the optimistic case are based on 1988 profit margins. Discussions with steel constultants, producers, and traders indicate that this assumption produces the best estimate of slab delivered price in 1985.

[c] Investment on continuous casting is to upgrade from 45 percent to 100 percent continuous casting.

Source: Results from net present value analysis.

per ton compared to a U.S. efficient integrated producer if one still existed in this region.

As a result, U.S. flat-rolled producers located in the East and the West can compete more effectively with Brazilian and South Korean imports of flat-rolled products, especially if customers are located nearer the plant site than the port. The same applies to the inland for both U.S. flat-rolled producers.

Table 6.27. U.S. International Competitiveness in the Flat-Rolled Market of Integrated Producers, Non-Integrated Producers Purchasing Slabs, and Importing Slab, Compared to Imports from Brazil and South Korea in 1988 (dollars per metric ton)[a]

	Total Cost at Delivered Region		
	Inland	East	West
Efficient Integrated[b]			
U.S. Inland	509.4	539.4	554.4
U.S. Coastal	545.2	515.2	560.2
Non-Integrated[bc]			
Efficient			
Inland	509.3	539.3	554.3
East	532.6	502.6	547.6
West	556.5	556.5	511.7
Inefficient			
Inland	532.8	562.8	577.8
East	555.6	525.6	570.6
West	580.2	580.2	535.2
Imports[d]			
Brazil	502.2	499.2	513.7
South Korea	501.5	497.5	495.0

Note: [a] Composite cost estimates throughout this chapter are based on a plant product mix of 30 percent hot-rolled sheet, 30 percent cold rolled sheet, and 40 percent galvanized sheet.

[b] Cost at the plant site.

[c] Importing Brazilian slabs.

[d] Considers trader sales commission of 3 percent and unloaded at the port added to total production cost. The unloading cost considered in different delivered regions are $4.5 per ton inland, $7.5 per ton in the East, and $10 per ton in the West.

Source: Calculated using cost estimation procedure of Appendix D and Table 6.16.

SUMMARY AND CONCLUSIONS

A major factor contributing to increased imports is the shift in the location of high-grade raw materials, low labor costs, and transferability of technology among countries. Overseas iron-ore mines in Brazil and Australia contain much higher grade iron ore than is available from North American mines, leading to lower raw-material costs, increased raw-material production, and increased semi-finished steel export opportunities for producers in those regions.

Scrap prices are expected to increase, and the difference in the production cost of molten iron between U.S. producers and offshore producers using high-grade iron ore at lower prices than the United States is expected to provide the incentive to rely on imported supplies of semi-finished steel. In addition, developments towards a larger market for semi-finished steel products are expected to continue as the industry decentralizes, moving to smaller, more geographically diverse production facilities. Reduction in scrap supplies creates the need for developing an efficient substitute that does not have the loss of metallic content occurring in pig-iron and merchant DRI use.

The survival of an efficient integrated plant does not depend on imported slab, but purchasing slab from Brazil can considerably improve profit margins. The survival of U.S. integrated plants with inefficient hot-end and efficient finishing-end is possible only when importing slab from the cheapest slab supplier, which, in all cases considered, was Brazil. Integrated producers with an efficient hot-end and an inefficient finishing-end have a higher probability of continued operations until their hot-end facilities fully recover the investments made in them than do integrated producers with the opposite characteristics. Both types of presently integrated firms might decide to operate as non-integrated producer slab purchasers until new iron-making technology is proven.

Therefore, importing slabs from Brazil may allow present finishing flat-rolled capacity to survive and may prevent increased flat-rolled product imports, especially in the West where domestic slab supplies are not economic. Expansion of finishing flat-rolled capacity is only profitable when investing in a green-field non-integrated producer purchasing imported slab from Brazil when a 6 percent discount rate is accepted and the producer behaves as an optimist.

In addition, imported slab from Brazil gives the U.S. efficient non-integrated producer the ability to compete with the most competitive foreign flat-rolled imports, especially in the East and in the West. The Western vulnerability to flat-rolled imports can be further reduced if efficient non-integrated slab purchasers expand their capacity.

Finally, should this assessment about slab imports be correct, production costs in new plants under consideration in the north of Brazil (Usimar), in Venezuela (CVG), and in Mexico (Sicarsta) are comparable to the present Brazilian slab production costs. The competitiveness of such new supplies is increased because estimated freight costs are about $10 per ton less than those for present ones.

7

Protectionism in Steel

Many countries, including the United States and the EEC, restrict steel imports to attenuate the effect of world-wide excess capacity on their domestic steel industries. The United States and the EEC restrict imported steel through quota agreements.

Import restrictions of all types raise domestic prices above free market levels and lessen competitive pressures on domestic producers, allowing inefficient firms to produce. Protection thus harms the consumers and the exporters of final products that contain steel. Given long-standing concerns about the inadequacy of competition in steel, such protection implies a major reversal of policy. As data in this book show, the protection neither restored profitability nor produced the modernization that protection was supposed to facilitate.

This chapter reviews the basic economics of protectionism, the history of U.S. import controls, and their application to slab imports.

THE ECONOMICS OF PROTECTIONISM

While this book is not directly concerned with the debate over protectionism, some basic points in the economics of protectionism should be noted. First, charges of unfair practices should be viewed suspiciously. Moreover, "unfairness," if it occurs, consists of a decision of the foreign producers or their governments to subsidize U.S. consumers. Protectionism thus deprives the U.S. economy of the benefits foreigners wish to provide.

The usual counter-argument is that the subsidies will last only until U.S. producers are driven out of business, and then prices will be raised above the levels that prevail under protection. This argument, in turn, has at least two defects. First, the ability to limit competition in this fashion is dubious. Second, it is not necessarily true that success at driving out competition is harmful on balance to U.S. consumers. The net effect depends upon the net present value of the benefits and costs to the U.S. economy. The benefits increase with the duration and magnitude of the price concessions made; the costs increase with the magnitude and duration of the subsequent price rises. For the U.S. economy to lose from a strategy of reducing competition, short, small concessions must lead to large, persistent price increases.

The idea of the need for temporary protection to permit long-run competitiveness is even more dubious. If competitiveness was clearly restorable, the steel industry could somehow finance the new investment needed. Protection is a third best approach. Presumably, any barrier to profitable investment would be some sort of capital-market imperfection. The best policy is to eliminate that imperfection. The second best approach is government lending, and generating the funds from profits increased by protection then comes in as third best.

The failure of 20 years of protection in steel to produce the promised turn-around should suffice to create doubts about whether the policy makes sense. The skepticism is reinforced by the record of world-wide policy failures in agriculture, retailing, coal mining, and many other industries.

International trade economists tend to argue that if trade is to be restricted, tariffs are the preferable policy. Tariffs leave the decision about who imports what from whom up to individuals in the market. Tariffs provide clear limits to the price impacts of protection, but the degree of import reduction produced is hard to predict. Quotas lead to predictable import levels but with difficult to predict price impacts. No clear-cut way exists to decide which system is preferable in terms of overall effects. This depends, among other things, upon how well either system can adapt to import levels that seem too high, or price rises that seem excessive (the argument is similar to that over alternative policies for pollution control as discussed by Baumol and Oates, 1988).

The preference for tariffs is based on their simplicity compared to quotas. The government must decide how to allocate the quotas. The theory of quotas shows that they are a right to buy at the import price and resell at the higher domestic price produced by import controls. At a minimum, bitter political battles may be waged to secure quotas and the profits they convey. Where political corruption prevails, bribes may be offered. Auctioning quotas could eliminate these frictions but is rarely done.

Asking importers or exporters voluntarily to limit imports involves even greater problems. Success requires the ability to form an effective cartel. The history of voluntary trade controls and other cartelization efforts suggests that competitive forces undermine the efforts. This has proven true in U.S. steel import programs.

U.S. TRADE BARRIERS AND THE IMPACT OF VER

From 1968 to 1986, three different trade-restricting mechanisms were imposed, each due to rising steel imports. First, from 1969 to 1974, Voluntary Restraints Agreements (VRA) were made with the EEC and Japan. Second, a Trigger Price Mechanism (TPM) prevailed from 1978 to 1982. In 1984, the government instructed the U.S. International Trade Commission (USITC) to negotiate Voluntary Exports Restraints

Agreements (VER) with countries that were principal exporters to the United States.

In response to the sharp increase in imports in 1968, the United States negotiated VRA with Japanese and European exporters limiting exports of steel to the United States to a target of 14 million tons in 1969 with gradual increases during subsequent years (Mueller, 1984). In 1969 and 1970, imports dropped to a level of 14 million and 13.3 million tons, respectively. However, Japanese and European producers began to supply the U.S. market with higher-valued products (American Iron and Steel Institute, various years), increasing the value of imports although tonnage declined. Estimates of the effects of the VRA on the U.S. steel industry (Crandall, 1981) were:

--an increase of capacity utilization by no more than 3 percent;
--an increase in prices by U.S. producers of somewhere between 1.2 percent and 3.5 percent; and,
--an increase in the average cost of imports for five major steel-mill products a minimum of 6.3 percent and a maximum of 8.3 percent.

After 1972, the VRA were not as necessary to preserve U.S. steel production as they were in 1969. World steel demand increased in the early 1970s. After a slight recession in 1970-71, the economies of the noncommunist industrialized countries experienced at the end of 1972 the most rapid upturn since 1958-59 (Crandall, 1981).

However, fears about over-heating of the economies in the major industrialized countries arose. Restrictive monetary and fiscal policies were applied by governments. These deflationary measures produced deep recessions that depressed steel demand (Acs, 1984).

During 1973-74, steel supply in Europe was tight and world market prices were higher than the prevailing domestically controlled prices. This price differential increased U.S. exports from an average of 2 million tons to a level of 5.8 million tons in 1974 (Mueller, 1984).

The 1968-74 period was one in which the tendency to ignore professed expectations of quid-pro-quo cost reductions was particularly pronounced. Little was done to increase efficiency and narrow the cost gap. Employment costs, for example, were allowed to increase at a rapid rate even before signing the 1974 Experimental Negotiating Agreement (ENA) (see Chapter 6).

Another error of the steel producers was to interpret a 3.9 million-ton drop in imports from 1974 to 1975 as indicating a long-term recovery in competitive position. However, imports rose again in response to a rapidly strengthening domestic steel market and several price increases by U.S. producers. Other producers, notably Canada and developing countries, were attracted by high prices in the U.S. steel market and began to ship larger quantities into the United States. In addition, European exporters attempted to regain the market share they held under the VRA.

Profits of the integrated steel producers were nearly zero in 1977. One contributing factor was that intensifying foreign competition precluded raising prices sufficiently to cover high U.S. costs (International Iron and

Steel Institute, 1980). Under the ENA, employment costs were correlated with import prices. As import prices rose, labor costs increased, pushing up the cost per domestic ton to levels far above those at which domestic firms could realize profits (Mueller, 1984).

An increase in imports from 12 million tons in 1975 to 19.3 million tons in 1977 renewed interest in barriers against steel imports. Intensive lobbying by the steel industry and the USW led to the formation of a Congressional "Steel Caucus." The government came under pressure to reduce the flow of imports but disagreed about how to negotiate new quota agreements. Instead, the government advised the industry to file anti-dumping petitions.

Therefore, the industry submitted a massive number of complaints which set off some retaliation from European exporters. The U.S. government devised the Trigger Price Mechanism (TPM) to avoid a trade war. The TPM was based on the estimated production cost for an efficient Japanese producer, plus an added 8 percent profit margin and freight. Unless import prices were above the trigger price, the country would be subject to investigation of whether it had engaged in dumping.

After the TPM was established, the anti-dumping petitions were withdrawn because the mechanism established a de facto minimum price for most steel products imported to the United States. From early 1978 to early 1979, the TPM worked quite satisfactorily for the domestic industry.

A joint industry-labor-government committee, which evaluated the system in June 1979, concluded that import tonnages and import penetration had declined since the inception of the TPM (Acs, 1984). However, increases in the trigger prices that favored the U.S. industry were almost exclusively the result of yen straightening against the dollar.

This benefit ceased after the exchange rate settled at a more stable level. In fact, from April to December 1979 the trigger price was reduced only slightly. Although the system brought little further gain, the majority of American steel firms still wanted to retain it. Its relative certainty seemed preferable to the unpredictability of the only then available alternative of large-scale anti-dumping actions. Nevertheless, in March 1980, USS filed an anti-dumping petition against the major EEC steel producers. In keeping with its understanding of the bargain, the government promptly ended the TPM.

After several months of negotiation, the government announced a new "Steel Plan" that raised trigger prices by 12 percent over their previous level, and the anti-dumping petition was withdrawn (Bright, 1984). This new truce between the government and the industry remained an uneasy arrangement. In 1981, the Economic Recovery Tax Act (ERTA) passed by Congress did not provide incentives for the steel industry to invest in modernizing capacity. As a result, the steel industry reacted by engaging in the activities discussed in Appendix A (Mueller,1984).

In January 1982, the new Administration refused to "self-initiate" a large number of trade cases being pushed by USS and several other major producers. Again the steel industry entered petitions for countervailing duties and other anti-dumping measures aimed at the European and Brazilian steel exporters (Mueller,1984).

In early 1980, when the TPM was temporarily suspended by the government, U.S. steel prices started to decline. Price cutting also arose in early 1982. The effects of the recessions of 1980 and of 82-1983 and renewed foreign competition spurred the integrated producers to call for more trade barriers (Mueller,1984).

The TPM and the preceding VRA did not lead to a substantial improvement in industry output, profitability, or investment. The Congressional Budget Office (CBO), using its steel model, indicated that the TPM did not have a significant effect on the price of imports or the demand for domestically produced steel.

However, the TPM still produced less import reduction than the domestic industry desired. After 1982, steel imports increased significantly, labor and management in the integrated steel industry began to lobby for tighter import restrictions, and on the eve of the 1984 presidential election, protectionism gained momentum.

Former steel workers at Armco's closed Houston Works initiated a petition drive in support of the industry's most recent protectionist efforts in Congress, the Fair Trade in Steel Act of 1984. The petition was placed on the final "wide-flange structural steel I-beam" produced by the now defunct Armco plant. The I-beam was carried on a 16-wheel tractor trailer bearing the sign "Foreign Steel Steals Jobs." The I-beam petition gathered thousands of signatures in 17 states before arriving in Washington, D.C., just as Reagan Administration officials began its testimony against the proposed legislation before the Trade Subcommittee of the House Ways and Means Committee. The 1984 bill represented a compromise effort between labor and management to protect steel and reverse the industry's disinvestment policies (Scheuerman, 1986).

In September 1984, President Reagan instructed the U.S. Trade Representative to negotiate Voluntary Export Restraint Agreements (VER) for a five-year period with the countries that would be the principal exporters of the steel to the United States. The current VER program represents a return to quantitative limits, albeit without a few of the exemptions allowed in the voluntary restraint agreements of the early 1970s. The VER are bilateral agreements covering imports from twenty countries. Export licenses, without which the product cannot enter the United States, are issued by foreign governments on the basis of the market-share projections provided by the United States. The agreed upon market shares are highly product-specific to limit mix changes. The program as a whole sets 186 limitations. Some countries managed to avoid signing agreements, but they are bound by significant political pressures to restrain from exploiting that status (U.S. Congressional Budget Office,1984).

In December 1984, agreements were made with nine large steel-exporting nations, but actual restraint agreements were not effective in the important cases until May 1985. Several factors contributed to delaying the impact of the import-restraint program. Many bilateral agreements with major exporters were not formally in place by midyear. Moreover, shipments surging from new suppliers were not subject to restraint.

Table 7.1 shows 1987 U.S. steel trade arrangements for finished steel. The first ten countries were the traditional exporters to the U.S. market that

Table 7.1. U.S. Steel Trade Agreements for Finished Steel in 1987

Market Shares Agreement[a]	Share of U.S. Apparent Consumption (percent)
Australia	0.25
Austria	0.21
Brazil	1.50
European Community	5.90
Finland	0.22
Japan	5.50
Mexico	0.41
South Africa[b]	0.00
South Korea	1.80
Spain[c]	0.70

Quotas Agreements[a] (metric tons)	Quantity
Czechoslovakia	36,280
East Germany	99,770
Hungary	30,838
Poland	81,630
Portugal[c]	36,280
Rumania	95,235
Trinidad and Tobago	66,665
Venezuela	180,584
Yugoslavia	22,856
People's Republic of China	61,676

Note: [a] Including semi-finished steel for all countries.
[b] Steel imports from South Africa in 1987 are reduced by the comprehensive Anti-Apartheid Act of 1986, which embargoes certain products.
[c] In 1988 had become part of the EEC.
Source: *World Steel Dynamics*, "Steel Strategist #14," 1987.

initially received their market share agreements. In 1984, South Africa had a market share of 0.42 percent, but in 1986 its steel exports were embargoed by the Anti-Apartheid Act. The other countries listed in the table reached an agreement after the first ten had been settled, but they agreed to specific amounts rather than a share of U.S. apparent consumption. While the agreements cover 20 countries, more than 30 countries are not covered by the quota. For example, Canada, the second largest single source of imported steel, is not covered by the quota because of its historically close economic ties with the U.S. markets.

All of these non-tariff mechanisms have functioned as price floors, increasing prices above the competitive level and reducing the quantity demanded, thus imposing economic inefficiencies. The historical review shows the failure of the two formal trade barriers to provide competitiveness in the U.S. steel markets and to provide the necessary profitability to modernize the integrated capacity. These mechanisms delay closing of inefficient capacity. The experience suggests that the policies were imposed to protect high wages.

In July 1984, the U.S. Congressional Budget Office released a study showing the results of its model. The model was run for a proposed 15 percent quota on steel imports. The projections generated by the model suggested that:

- the average price of steel consumed in the United States would increase by about 10 percent;
- the domestic steel prices would be roughly 3 percent higher in the first year of the quota and roughly 7 percent higher in the fifth year;
- the price of imported steel would be roughly 34 percent higher in the first year of the quota and roughly 24 percent higher in the fifth year;
- U.S. steel consumption would decrease by between 4 and 5 percent;
- domestic steel shipments would increase by 6 percent;
- steel industry employment would increase by 6 percent to 8 percent;
- the quota agreement would transfer between $2 billion (in the first year) and $5 billion (in the last year) to the domestic steel industry in the form of pretax profits; and,
- the agreement would transfer about $2 billion annually to the foreign steel producers, presuming the U.S. government did not seek to capture this sum through auctioning off import licenses or similar measures.

In short, this quota, like most protection, would benefit domestic producers and workers at the expense of steel-consuming industries and their customers. The steel-consuming industries would increase prices. The agreement would be particularly injurious to the steel-consuming industries that face international competition.

The CBO estimated that quotas would cause efficiency losses of approximately $1 billion per year. When these are added to the income transfers that occur as a result of the quota, it is suggested that a 15 percent steel quota would cost U.S. consumers between $4 billion to $6 billion (in 1983 dollars) during each of the five years the quota was in effect. These costs would increase over the duration of the quota.

SLAB IMPORT RESTRICTION AND ITS EFFECTS ON DOMESTIC STEEL PRODUCERS

Since 1984, imports of semi-finished steel rose significantly in the U.S. market. Although included in the overall quota agreement, these imports are controlled by specific arrangements that set limits by restricting physical quantities and not by the share of apparent consumption. Table 7.2 shows the semi-finished steel agreement for individual countries. Brazil is the largest single supplier of semi-finished steel, followed by Japan, Mexico, and Venezuela.

Protectionism in Steel

Table 7.2. U.S. Steel Trade Arrangements for Semi-Finished Steel in 1987

Quotas Agreements	Quantity (metric tons)
European Community	743,740
Japan	90,700
Brazil	634,900
South Korea	45,350
South Africa[a]	0
Australia	45,350
Mexico	90,700
Spain[b]	45,350
Finland	13,605
Venezuela	54,420
Total	1,764,115

Note:[a] Steel imports from South Africa in 1987 are reduced by the comprehensive Anti-Apartheid Act of 1986 which embargoes certain steel products.
[b] In 1988 had become part of the EEC.
Source: *World Steel Dynamics*, "Steel Strategist #14," 1987.

The restriction of imported semi-finished steel introduced two side-effects. First, the semi-finished import constraint expectedly lessened competition and pushed up the prices of semi-finished products. Second, it undermined possible increases of U.S. steel exports and discouraged joint-ventures with foreign investors.

The Effect of VER and Short Supply Mechanism

Imports of semi-finished steel shapes are constrained by the VER program at 1.8 million tons per year; however, additional amounts can be obtained through the so-called "short supply" mechanism provided in every bilateral agreement signed by the United States. To secure more imported slab, U.S. customers must make a strong technical and economic case for a special allocation and develop support for their position, which is then submitted to the U.S. Department of Commerce (DOC). The processing of the short supply request takes an average of two months from when it is submitted until it is approved.

The U.S. requestor must include as a minimum the following:

--exact specifications of the product;
--a list of all U.S. producers from whom purchases of slabs were made in the past three years;
--a list of all U.S. and foreign producers of the product that have refused to sell the product in the last three years, indicating the reason for the refusal;

--a list of all offers to sell the product by U.S. and foreign producers in the last three years, offers that have been rejected by the requestor, indicating the proposed terms of sale and the reason for the rejection;

--a list of other U.S. consumers of the product including supporting statements indicating all U.S. and foreign producers of the product that have refused to sell the product in the last three years, indicating the reason for refusal;

--the requestor's estimate of the operating rates of U.S. producers of the product and an explanation as to why these figures demonstrate the existence of short supply; and,

--the information available to the requestor, such as that pertaining to allocation or delivery times, that the requestor believes demonstrates the existence of the short supply (Barringer, 1986).

Short supply requests are approved if domestic supplies are not considered to be available. The price of domestic supplies is not relevant, and the approvals for short supply imports are good only for a three-month period. The verification of the existence of domestic supplies and the duration of the short supply approval imposes constraints on other U.S. steel producers.

In 1987-88, a slab deficit became evident as the operating rates of U.S. steel producers varied between 80 percent and 95 percent. Imports of semi-finished steel increased to the point that the VER ceilings limited imports. Imports of semi-finished steel shapes reached the ceilings under various agreements.

High levels of demand for finished products throughout 1987 and 1988 led to numerous filings by domestic firms for semi-finished steel under short supply provisions. In both years, the DOC, which administers the program, found that "extraordinary circumstances" existed and approved extra semi-finished steel imports.

Table 7.3 shows the short supply requests and the quantity approved. Of the 1.9 million tons requested until May, 1988, only 41 percent was approved. In 1987, DOC approved 0.3 million tons of short supply imports and through July, 1988 DOC had approved 0.5 million tons.

Actual imports under short supply waivers were perhaps only half of what was approved, due to the tight supply conditions existing worldwide for semi-finished steel. Therefore, delivery scheduling problems arose because approvals of short supply imports are good only for three months. Industry frustration with the approval process prompted a group of ten steel caucus members to request that the General Accounting Office (GAO) review the administration of the short supply procedure, with special attention to means of speeding up the approval process.

In the requests for short supply waivers, companies cited various justifications. A general lack of domestically available material was the overriding cause, but special cases also caused requests. The six-month strike of USS generated several requests by the plants purchasing its slab, such as Tuscaloosa. Some of these requests were withdrawn with the resumption of production by USS. A number of companies also cited equipment shutdowns, failures or production disruptions caused by

Table 7.3. Short Supply Requests for Slab by U.S. Steel Mills in 1987 and 1988

Company	Original Date	Period	Quantity of Original Request (tons)	Amended Date
California Steel	11/21/86	1987	90,700	11/2/87
				12/8/87
	1/6/88	1&2q 88	90,700	1/29/88
	3&4q 88			3/9/88
Gulf States	8/28/87	4q 87	67,653	
	1/25/88	1&2q 88	57,595	2/1/88
Lone Star Steel	11/13/86	1987	370,963	
	12/4/87	1&2q 88	204,982	
		3q 88		4/1/88
Lukens	10/28/87	1q 88	20,408	
	3/23/88	3&4q 88	45,350	
National Steel	9/29/86	87,1q88	217,680	
Sharon Steel	2/26/87	1987	208,610	7/29/87
	12/23/87	1988	85,880	
Tuscaloosa Steel	4/8/88	2,3,4q88	375,251	
Weirton Steel	12/14/87	4q 88	34,446	
		3q 88		3/30/88
Total			1,870,238	

Source: U.S. International Trade Commission, *Annual Survey Concerning Competitive Conditions in the Steel Industry and Industry Efforts to Adjust and Modernize*, Report 2115, 1988.

installing new technology either in their own facilities or their suppliers, as the cause of their requests. Other reasons have ranged from the embargo of South African slabs to the construction of new rolling mills prior to the completion of furnaces at the facility to feed the mills.

Another issue concerning short supply requests was the price of domestic semi-finished steel. The DOC cannot approve offshore purchases if a domestic company agrees to supply the requester with semi-finished steel, regardless of price. Although the DOC contended that price could be a consideration if it is prohibitive, it has not made that finding since the beginning of the program in 1984.

In 1987, Lukens Steel stated that domestic prices (USS) were 15 percent to 20 percent above the price of foreign slab. As a result, plants purchasing domestic slabs have a very slim profit margin, if any. Lone Star Steel and

Table 7.3. (continued)

Amended Quantity (total tons)	Approval Date	Amount Approved (tons)	Comments
	11/24/87	27,210	No melting facilities; affected by
	12/31/87	45,350	USS strike, and South African
16,326			supplies cut off by embargo
190,470	4/29/88	66,520	
	11/24/87	19,047	Excess rolling capacity
60,497	3/2/88	60,497	Had furnace problems
	4/27/87	5,597	Shut down blast furnaces,BOFs,
	5/15/87	19,455	ore mines, and coking facilities
	6/23/87	23,802	
	8/13/87	40,815	
	12/7/87	40,815	
	3/18/88	90,700	
184,575	4/29/88	142,853	
	3/18/87		Long standing as semi buyer
	4/29/88	11,338	and excess rolling capacity
			Short term needs while upgrading
			refining &casting facilities
96,686	10/13/87	40,768	Furnace problems
	4/29/88	37,388	Short on Blast Furnace Capacity
	5/16/88	62,583	No melting facilities
			Critical specification material
	4/29/88	34,466	Considering melt shop improvement
		769,193	

Gulf States Steel similarly complained that USS prices were substantially above the price of both foreign and domestic slabs from other producers. Lone Star stated that the prices of end products had gone up 17 percent over the previous year, but the slab price was 57 percent higher. Also, Lone Star said the part of the short supply request denied by the DOC was to be supplied by USS, but USS was unable to meet its commitment. The data in Table 7.3 support these contentions.

The VER and the limited relief given under the short supply mechanism thus pushed up steel prices. The slab price and the time required to approve short supply requests were the main influences. The excess demand for slab, denoted by the rise in slab prices, would eventually raise the prices of flat-rolled products because it generates excess demand for finished products. Finished product excess demand would be eliminated totally by rising prices, restricting consumption, and inducing increased supply from marginal facilities.

Joint-Ventures and the Redundancy of the
Semi-Finished Steel Imports Restrictions

The VER and the possibility of its extension also have undermined any possibility of joint-ventures and of establishing long-term contracts with foreign producers. A domestic producer lacks incentives to participate in joint-ventures, partial ownership of foreign slab plants, or long-term contracts with foreign producers if the slabs cannot enter the United States. These joint-ventures, foreign ownership of foreign slab plants, and long-term contracts can provide the U.S. producers with a means to secure foreign supplies of semi-finished steel at lower prices than domestic spot prices.

An example of business opportunities from which U.S. producers have been excluded are drawback operations. These operations consist of importing a semi-finished product to be processed and then exported. This type of operation would allow the U.S. profitably to raise its steel exports revenues by the re-processing of semi-finished steel shapes into finished products if the revenues are higher than the costs involved in purchasing and re-processing.

If such opportunities existed, higher finishing facilities operating rates would prevail, new jobs would be created, and the risk of hiring new employees would decline. In 1988, employees were working extra hours to meet demand when operating rates were above 80 percent because companies were afraid of hiring (U.S. International Trade Commission, 1988a).

The involvement of U.S. steel producers in partnerships with foreign producers has been undermined by the political pressure of the steelworkers union. This may be against the long-run interests of the workers. Their implicit belief is that it is possible to maintain an integrated steel industry with employment in hot-end operations. An alternate possibility is that the level of protection that is being provided will lead to the demise, not only of hot-end operations, but of finishing facilities that could otherwise survive. Thus, steel industry employment might ultimately be less under continued protection than under free trade.

Experience shows that steel jobs are lost if the ability to purchase slab is not available. First, the Kaiser closure in 1984 resulted in unemployment of steelworkers, and 700 of these jobs were recovered with the start up of operations of CSI, which has no steelmaking capacity. Second, in 1983-84, the union opposed the USS import of slab from British Steel and the purchase of Brazilian slab by Wheeling-Pittsburgh. The strong opposition of the steelworkers stopped the two negotiations. This probably contributed to job losses and plant closings Third, the ability to purchase slab created new jobs at Tuscaloosa, which buys part of its supplies from foreign producers and from USS. In addition, the profit prospects of buying imported slabs caused proposals of new ventures such as the USS-Posco and Heidtman, which would create more jobs.

SUMMARY AND CONCLUSIONS

As data presented elsewhere in this book indicate, limitations on semi-finished steel were ineffective in restoring the competitiveness of U.S. steel producers, improving their profitability, and enabling them to modernize.

The VER semi-finished import quotas were not sufficient to reduce the slab deficit in 1987 and 1988, as producers asked for "short supplies." Therefore, VER limitations of semi-finished steel imports generates economic inefficiencies and limits the ability of flat-rolled producers to increase their profitability.

The ability of U.S. steel producers to compete internationally may be endangered because they cannot take advantage of the business opportunities to which their competitors have access. In addition, the VER make it unattractive to pursue joint-ventures, long-term contracts, or direct investment with semi-finished producers overseas, such as Brazil. Semi-finished steel not imported by the United States most certainly will be processed into finished products in other countries, such as Turkey and the Philippines, and these finished products may end up being sold on the U.S. market. In short, steel protectionism may ultimately harm the steel industry as well as the rest of the economy. It is hoped that eventually the constraints will be removed.

8

Summary and Conclusions

The world steel industry has been marked by excessive capacity and growing international trade. The combination of these two factors produced a more competitive environment in international and domestic steel markets. This more competitive environment has been influenced by the increasing participation of non-traditional steel producers such as South Korea and Brazil. These newer participants have lower costs than traditional steel producers, which have introduced downward pressures on steel prices. To reduce the impact of a more competitive international steel trade on individual markets, traditional steel-producing countries--namely industrialized countries--engaged in steel import restraints. This protection was not sufficient to obviate adjustment, and steel industries in most industrialized countries were forced to restructure their capacities.

The United States is one of the countries affected by reduced competitiveness and engaged in trade restrictions as net imports increased. The difficulties predominantly affected integrated steel producers. They lost markets to the growing U. S. mini-mill producers and to imported steel. Production costs of U.S. integrated steel producers exceeded those of other major steel-producing countries and, for some products, those of mini-mills. Three major variables have influenced this competitive condition: labor, iron ore, and exchange rates.

However, the efforts of integrated producers to cut labor, iron ore, and other costs, as well as the 1985 devaluation of the dollar against major strong currencies, has made U.S. integrated producers lower cost than those in Japan and West Germany, two of the largest steel exporters to the U.S. market. On the other hand, U.S. integrated steel producers still are not as competitive as those of South Korea and Brazil.

The low demand for steel associated with high production costs produced poor financial results for integrated steel producers from 1982 to 1986. During this period, the U.S. steel industry accumulated losses of $12 billion, most of which were due to the losses of integrated producers. The financial difficulties of integrated producers resulted in bankruptcies and in the restructuring of some steel producers. At the same time, foreign investors and American entrepreneurs were attracted by individual projects with a potential of high return on invested capital, such as continuous casting, electro-galvanizing, and buying closed finishing facilities, or building finishing facilities to re-process semi-finished steel.

Restructuring of the U.S. integrated steel industry involved partial or total plant closures, decreased employment, and a major production cost-reduction race, which became more evident since 1982. Decreasing steel-industry employment strongly influenced the imposition of import restrictions. On the other hand, the partial and total plant closures have introduced imbalances in the producing capacity of integrated producers.

The ability to produce steel slabs is now less than the capacity to finish The calculations shown here on U.S. steel industry production capabilities confirm the existence of this disparity. Practical evidence is provided by increasing imports of carbon-steel slabs. This slab deficiency is likely to worsen because most flat-rolled steel product producers are avoiding major investment in iron-making facilities. The profitability of upgrading inefficient hot-end facilities is questionable, and environmental regulations create further uncertainties about profitability. Prospects that better steelmaking technologies will be perfected add to the risk. Investments in facilities to produce the higher- quality product that consumers now demand seem more profitable than investments in hot-end facilities.

Restructuring of steel capacity and the high risk of investing in conventional technology, stimulated the growth of international trade of semi-finished steel, especially from countries specializing in semi-finished steel production at low costs. International trade of semi-finished steel permitted countries to produce finished steel products using cheaper inputs than could be produced at green-field plants in the United States. Trade in semi-finished steel also has allowed some countries, such as Turkey and the Philippines, to increase their participation in the international trade of finished steel products.

U.S. domestic supplies of semi-finished steel declined, and the United States became the largest importer of semi-finished steel. As of 1987, these imports represent only 3 percent of the crude-steel production. However, a potential does exist for substantial growth. Several million tons of low-cost slab producing capacity is available in Brazil, South Korea, and other newcomers to the steel industry.

Restructuring of U.S. integrated steel producers also caused differences between the East and the West. The West experienced a heavy penetration of flat-rolled product imports that fell as imports of semi-finished steel increased. The dependency on semi-finished steel of the West is larger than that in the East and for the United States overall. This has occurred because the West has only one integrated producer and it has no continuous casting facilities and a limited ability to supply slabs.

To overcome the present limitations of semi-finished steel imports, domestic producers have engaged in the re-processing of hot-rolled coils. This occurrence is noticeable in the overall U.S. market and, more specifically, in the West, where USS-Posco and LTV's Hennepin facilities are operating. Both producers are re-processing hot-rolled coils to produce other finished flat-rolled products.

Summary and Conclusions

The United States imposed three types of trade barriers against imported steel. The one prevailing in 1989, VER, was the first to impose specific limitations on semi-finished steel. The limitations on semi-finished steel have been in effect since 1985, when the slab deficiency was not apparent because of the low operating rate of U.S. industry. However, as the operating rates improved in 1987-88, a slab deficiency emerged, and short supply requests were filled for extra imports of slabs. These requests were not filled as rapidly or as fully as flat-rolled product producers desired. Thus, producers' complaints have arisen. These include high domestic prices of slabs, the denial of short supply requests based on the existence of domestic supplies that later were proved to be nonexistent, the complexity of filling short supply requests, and the problem of scheduling deliveries and negotiating prices within the period for which the import grant is valid. VER steel quotas and the administration of the short-supply request process have introduced economic inefficiencies by creating an excess demand of slab that will eventually result in rising finished steel prices.

Restrictions on imports of semi-finished steel also limited the U.S. producers' ability to explore long-term contracts and joint-ventures with foreign slab suppliers that would secure imported slab. In addition, the controls discourage U.S. flat-rolled product producers from engaging in operations that would boost U.S. steel exports and raise operating rates. The profitability of slab purchasing for further processing into finished products is greatest when slabs are imported. Imported slab has reduced the cost of entering the flat-rolled product market. In addition, integrated producers may be better off closing their hot-end facilities and processing slab than in producing steel on an integrated basis.

Producing from imported slabs may justify investing in a finishing capacity that would be unprofitable if domestically produced slabs are used. Importing of slab may on balance increase employment in the steel industry. Given the limits on slab supply these imports will displace only part of domestic hot metal production. No claim is being made here that total dependence on slab imports would be profitable in the 1990s.

Therefore, the trade limitations imposed may have proven contrary to the VER's goals of improving the international competitiveness of the U.S. producers. The controls are not producing the promised revival of U.S. integrated steel producers. More importantly in the steel industry, present policy may undermine the hidden goals of maintaining jobs in the steel industry and raising operating rates. With review of the VER agreements scheduled for 1991, it seems reasonable at least to eliminate the limitations on semi-finished steel imports. This import should be regarded as a raw material that makes it possible for U.S. steel producers to restructure and become internationally competitive. This is critical in the Western United States, which is more vulnerable to flat-rolled product imports. Moreover, liberalizing restrictions on steel-slab imports would increase the chances of lowering net direct and indirect steel imports and raising the attractiveness of producing finished steel products in the short and long run.

It should be evident from this book that the changing economics of steelmaking has made steel production move closer to raw materials and further from markets. Also, the industrial structure of the steel industry is becoming less concerned with the development of technology. As a consequence, Stigler's proposition that broadening markets tend to lessen vertical integration applies to the steel industry because of the increased linkages of previously geographically isolated markets.

APPENDIX A

The Financial Situation of U.S. Integrated Producers and the Implications for Investment

This appendix presents selected data on the financial performance of the U.S. steel industry. First, the financial situation is reviewed. Then, devices, including joint ventures with foreign producers, to finance investment are reported. Finally, data on bankruptcies and their consequences are explored.

FINANCIAL PERFORMANCE

Performance is so poor that every indicator for integrated producers is unfavorable. The prevalence of losses makes many conventional measures negative. From 1959 to 1981, the U.S. steel industry earned positive returns on equity (net income per shareholders' equity); however until 1987 the industry endured losses. From 1982 to 1986, the U.S. steel industry was one of the most unprofitable industries in the United States, with combined net industry losses totalling $12 billion. When capacity utilization reached 80 percent in 1987, a positive net income returned.

The financial problems of the U.S. steel industry predominantly affect integrated producers. In the period analyzed, the six largest integrated producers accounted for more than 56 percent of the U.S. market. The losses that the six largest integrated producers endured from 1982 to 1986 constituted 70 percent of total U.S. steel industry losses (Chase Econometrics, 1987). Manufacturing industry, excluding steel, had positive returns (*MBM*, 1986). In 1982-83, when operating rates were the lowest in the history of the U.S. steel industry, the losses of the six largest integrated producers totaled $5.5 billion and accounted for 97 percent of the losses of the U.S. steel industry (Chase Econometrics, 1987). The U.S. steel industry's profit margin (the ratio of net income to net sales revenue)

Table A.1. Percent of Return on Sales for the Six Largest U.S. Integrated Steel Producers and Nucor--1973 to 1986

Year	Armco	Bethlehem	Inland	LTV	National	USS	Nucor
1973	4.50	5.00	4.50	1.20	4.70	4.70	5.30
1974	6.40	6.40	6.00	2.30	6.50	6.90	6.00
1975	3.80	4.90	3.90	0.30	2.60	6.80	6.30
1976	3.90	3.20	4.40	0.70	2.90	4.80	4.90
1977	3.40	-8.30	3.30	-1.10	1.90	1.40	5.90
1978	5.20	3.60	4.90	0.80	3.00	2.20	8.40
1979	4.90	3.90	3.60	2.20	3.00	-2.30	9.90
1980	4.70	1.80	0.90	1.60	2.30	0.40	9.30
1981	4.30	2.90	1.50	5.10	2.10	7.70	6.40
1982	-6.40	-27.90	4.20	-3.20	-15.20	-1.90	4.60
1983	-16.20	-3.30	-3.80	-3.90	-5.20	-6.60	5.10
1984	-7.30	-2.10	-1.30	-5.40	0.60	2.60	7.70
1985	1.50	-3.80	-5.90	-8.80	-0.60	2.10	6.70
1986	-17.90	-3.50	0.60	-44.70	-1.00	-12.30	6.10

Source: *World Steel Dynamics*, "Steel Strategist #14," 1987.

deteriorated to a negative 12 percent in 1982 and remained negative until 1986 (AISI, 1987).

Table A.1 shows the profit margin, or return on sales, for the six largest integrated producers and Nucor from 1973 to 1986. During this period, the return on sales deteriorated for most of the six largest integrated producers. Since Nucor may be able to compete in the flat-rolled market by the development of thin-slab casting, it is used as a benchmark for comparison with the six largest integrated producers performance. Its return on sales increased in the period examined here.

From 1982 to 1986, the capacity of the six largest integrated firms fell 23 percent, the operating rate rose, and losses fell (Table A.2). In contrast, from 1973 to 1986, Nucor increased its return on sales and improved its financial performance. Although the operating rate of Nucor declined to 62 percent in 1982, Nucor showed a positive net income of $22 million (Table A.2).

A mini-mill is less subject to losses because it requires less investment per unit than do integrated producers. High investment costs are only a problem for companies whose demand prospects are imperiled. Unexpected, persistent declines in demand can lead to prices close to variable cost. High fixed costs denote substantial investments on which the required returns will not be earned when prices are close to variable costs. The main practical implication of such a problem is that facilities earning inadequate returns should continue operating only as long as variable costs (in the economic sense that includes maintenance costs) are covered.

The financial structure of the firm mainly affects the form of the adjustment. Equity investors automatically incur the losses. Those supplying bond financing will be repaid if incomes are sufficient. High debt ratios increase the commitment to debt service and the difficulty of repayment. Given the ability to default or declare bankruptcy, this commitment is not actually much more binding than that to stockholders.

Financial Situation of U.S. Producers

Table A.2. Net Income and Operating Rate of the Six U.S. Largest Integrated Steel Producers and Nucor--1981 to 1987

	Net Income (million dollars)		Operating Rate (percent)	
Year	Six Largest	Nucor	Six Largest	Nucor
1981	1,734.00	35.00	72.7	83.2
1982	-2,999.00	22.00	44.1	62.5
1983	-2,457.00	28.00	50.9	74.4
1984	-313.00	45.00	63.6	81.7
1985	-722.00	58.00	67.5	89.9
1986	-1,876.70	n.a.	59.7	89.9
1987	896.70	n.a.	64.9	86.8

Source: Compiled from *World Steel Dynamics*, "Steel Strategist #14," 1987.

The share of fixed costs for mini-mills' cost of production is less than 20 percent, while for integrated producers, fixed cost accounts from 25 percent to 35 percent of the total cost (*World Steel Dynamics*, 1986).

In 1987, U.S. steel producers earned a net income of $1 billion and a 3.8 percent return on sales. All integrated producers showed improved profitability. Although the recovery of demand contributes to higher prices, the improvement of financial results also was influenced by the lower value of the dollar, cost cutting, and a better competitive position of steel consumers in both the domestic and international markets.

Another problem in interpreting the data is their distortion through the effects of tax laws that created investment incentives particularly useful to the steel industry. The true after-tax profitability of the industry depends on company-specific tax situations on which data are unavailable. Given the losses on equity income and these tax effects, measures of return on total investment are used to provide an alternative performance appraisal. The measures used are the ratio of net income plus interest to total assets (PROA) for return on assets and the ratio of net income plus interest to long-term liabilities plus shareholders equity (PROI) for the return on invested capital.

Table A.3 shows the evolution of PROA and PROI for the U.S. steel industry from 1979 to 1987. The drop in the industry's net income in 1982, 1983, 1985, and 1986 resulted in negative PROA and PROI. The companies' revenue growth in 1984, in which steel mill products prices rose, produced a positive PROA (3 percent) and PROI (4 percent). By 1987, both financial indicators recovered to the 1981 level, resulting in a PROA of 7 percent and a PROI of 9 percent.

Non-integrated producers, primarily mini-mills, achieved a substantially higher return on capital over the 1979-87 period than did major integrated producers. During the 1976-80 period, non-integrated mini-mills achieved on average a return on investment above 12 percent, while integrated producers secured a return just over 5 percent (*American Metal Market*, 1987).

Table A.3. Indicators of U.S. Steel Industry Return on Investment and Assets and the Capital Intensity--1979 to 1987

Year	PROA[a] (percent)	PROI[b] (percent)	Gross Capital/Sales Ratio[c] (times)	Net Capital/Sales Ratio[d] (times)
1979	4.42	5.62	1.12	2.40
1980	3.84	4.88	0.99	2.10
1981	6.75	8.56	1.11	2.37
1982	-9.99	-12.35	0.73	1.58
1983	-6.34	-8.15	0.71	1.54
1984	2.84	3.63	0.85	1.81
1985	-4.68	-6.08	0.83	1.80
1986	-16.76	-20.76	0.76	1.85
1987	7.05	9.17	0.85	2.06

Note:[a] PROA is the ratio of net income plus interest expenses to total assets.

[b] PROI is the ratio of net income plus interest expenses to long term liabilities plus shareholders' equity.

[c] Gross capital sales is the ratio of sales revenue to the value of property, plant, and equipment.

[d] Net capital sales is the ratio of sales revenue to the *net* value of property, plant, and equipment.

Source: Compiled from American Iron and Steel Institute, *Annual Statistical Report*, Various Years.

In the years that followed, mini-mills continued to have higher profit rates than did integrated producers. In the 1982-86 period, most integrated producers reported a negative return on invested capital and on assets. The rate of return for integrated producers reporting positive net income during this period was lower than the all-industry average and the profit rates of mini-mills. In the 1982-83 period, mini-mills also ran losses. Although mini-mill profitability was restored, losses by integrated producers continued through 1986 (*World Steel Dynamics*, 1988).

This decline in the ratio of investment to sales resulted partially from a drop in the fixed assets of the U.S. steel industry from $36 billion to $33 billion (AISI, 1987) (Table A.3). This decline was more influential than the decline of sales revenue from $40 billion to $26 billion (AISI, 1987). One cause was the increased contribution to sales by mini-mills; as noted, they have lower capital/sales ratios than integrated producers. Another influence is that the integrated producers reduced capital/output ratios by closing facilities.

The ratio of net capital to sales, which is based on the depreciated rather than the initial cost of fixed assets, also moved in a fashion reflecting the impact of declining sales revenues from 1982 to 1986. Net capital intensity

Financial Situation of U.S. Producers

recovered in 1987 to a level slightly below the 1979-81 level because of the influence of capital reduction on the steel producers.

The stock market began to anticipate problems in the U. S. steel industry long before 1982. In 1959, the ratio of market to book value peaked at 1.6 and then declined below unity (Crandall, 1981). Steel imports, which became significant in the early 1960s, may have served as a warning signal.

A 1984 Standard & Poor's Corporation report indicated that the stocks of 36 major steel producers surveyed traded at 73 percent of their book value (*Wall Street Journal*, 1985). Of the six largest integrated producers, Bethlehem traded at 88 percent of its book value; Armco, at 73 percent; the other four, at below 73 percent (Table A.4). Therefore, the combined $8 billion stock-market value of the six largest steel producers was less than the cost of building just one completely new large steel plant. In contrast the 1984 market value of Nucor was 207 percent of its book value (Table A.4).

Market values for individual steel plants, on a per ton of capacity basis, can vary widely depending upon such factors as location, product line, equipment condition, extent and nature of long-term liabilities, and special considerations--such as economic dependence of suppliers and community groups.

In several cases, steel plants have been purchased for little more than their real estate and scrap metal values. The average selling price, in the period 1984-87, for a steel plant in the United States ranged from $50 to $60 per ton of crude-steel capacity. This is equivalent to selling steel plants at a discount of 60 percent to 90 percent of the book value of total assets. For integrated plants, selling prices have ranged from $10 to $15 per ton for marginal idled capacity to $90 to $100 per ton for competitive operating capacity (Chase Econometrics, 1987).

From 1976 to 1986, steel share prices fell by more than half while the Standard and Poor's Index of 400 stocks more than doubled. This presumably reflected pessimism about prospects. The better the prospects that earnings will remain stable or even grow, the more valuable the company, and the higher the stock price. A falling price suggests that prospects are unfavorable.

Table A.5 shows the performance of earnings per share and stock price per share for the U.S. steel industry and the Standard & Poor 400 (SP 400) from 1973 to 1986. Earnings per share of the U.S. steel industry exceeded those of the general stock market only in 1974-75. In other years, the earnings per share in steel were below the market average and were negative since 1982.

The P/E ratio (the ratio of market price per share and the net income per share) is another broad, widely used performance indicator. Investors are willing to pay more per dollar of current income if the growth prospects are greater. Comparisons of price earnings ratios for profitable companies indicate differences in growth expectations. The P/E ratio of the U.S. steel industry peaked in 1977. The rebound in 1979 left the ratio below the 1977 level.

P/E ratios for the overall stock market were consistently positive from 1973 to 1986 and increased since 1979. Given that the steel industry has run consistent losses since 1982, its price earnings ratios were negative

Table A.4. Market Value of Common Stocks Compared to the Share Asset Value of the Six U.S. Largest Integrated Steel Producers and Nucor in 1984

Company	Market Value per Common Share (dollars)	Book Value per Common Share (dollars)	Market Value as Percent of Book (percent)
Armco	10.00	13.78	73
Bethlehem	17.63	20.03	88
Inland	24.38	42.14	58
LTV	8.50	12.87	66
National	28.50	42.71	67
USS	29.00	45.32	64
Nucor	43.88	21.16	207

Source: *The Wall Street Journal*, August 19, 1985.

(Table A.5). The existence of a positive price for the companies thus reflects belief that some asset value can be secured through recovery, liquidation, or some other changes.

Tables A.6, A.7, and A.8 show the earnings per share, stock prices, and the P/E ratios, respectively, for the six largest integrated producers and Nucor. The 1986 stock prices of the six largest integrated producers were at or below their 1973 levels. With a few exceptions, earnings per share and the P/E ratios were negative from 1982 through 1986.

In contrast, the stock price of Nucor increased from $2.9 to $41.0 per share, and its earnings per share increased from $0.51 to $2.17 per share. As a result, the P/E ratio of Nucor was positive and growing throughout the period analyzed.

Table A.9 shows that Moody's bond ratings for the six largest steel producers have declined since 1982. The ratings are an indication of confidence in the company's ability to repay. Interest rates on company-issued debt tend to be higher where bond ratings are lower. Bond ratings are based on expectations about a company's performance, so lower bond ratings represent a negative appraisal of the industry's financial prospects. The poor Moody's ratings of the six largest integrated producers reflect persistent losses and high debt levels. Presumably Moody's doubts whether improved profitability will continue. Over the 1979-87 period, both the liquidity and leverage of the U.S. steel industry deteriorated. Liquidity is indicated by the current ratio (the ratio of current assets to current liabilities) and the acid test (the ratio of monetary assets to current liabilities). The ratios dropped from 1.68 to 1.52 and 0.21 to 0.12, respectively (Table A.10). The greater decline in the acid test indicates that the industry's short-term ability to repay debt deteriorated. The decline in liquidity involved reduced working capital. Working capital fell from $4.2 billion to $1.4 billion from 1979 to 1985 but recovered to $2.65 billion in 1987 (Table A.10).

Leverage as indicated by the debt capitalization ratio (the ratio of non-current liabilities to non-current liabilities plus shareholder's equity) increased from 38 percent to 82 percent during the same period (Table

Financial Situation of U.S. Producers

Table A.5. Earnings per Share and Stock Prices for the U.S. Steel Industry and the U.S. Standard & Poor's 400--1973 to 1986

Year	Earnings per Share		Stock Price		P/E Ratios	
	SP 400	Steel	SP 400	Steel	SP 400	Steel
1973	8.88	7.16	120.40	41.77	13.56	5.83
1974	9.68	14.36	92.90	46.11	9.60	3.21
1975	8.28	9.16	96.60	58.64	11.67	6.40
1976	10.68	7.44	114.50	73.82	10.72	9.92
1977	11.59	2.32	108.40	56.41	9.35	24.31
1978	13.12	7.57	106.20	45.05	8.09	5.95
1979	16.21	4.11	114.80	43.49	7.08	10.58
1980	16.13	7.14	134.50	44.39	8.34	6.22
1981	16.70	13.57	144.20	50.96	8.63	3.76
1982	13.21	-14.46	133.60	34.81	10.11	-2.41
1983	14.73	-17.76	180.50	42.02	12.25	-2.37
1984	17.95	-2.19	181.30	40.78	10.10	-18.62
1985	15.24	-5.23	209.60	36.61	13.75	-7.00
1986	14.41	-10.50	263.30	31.47	18.27	-3.00

Source: *World Steel Dynamics*, "Steel Strategist #14," 1987.

A.10). The six largest integrated producers became increasingly more leveraged (Table A.11) and reduced their liquidity in the years considered (USITC, 1988). In contrast, Nucor became less leveraged and increased its liquidity.

The financial situation of integrated steel producers made them more dependent on debt capital. With negative net income, internal sources of finance were necessarily less than depreciation allowances. (Internal funds are essentially the sum of net profits and depreciation allowances.) Internal sources of financing have been nonexistent since the early 1980s. The losses also precluded new issues of equity capital on terms that the companies would find attractive because of the high risk and low potential of returns of investment in steel-making capacity. This dependence on debt capital and deterioration of liquidity made the U.S. integrated producers more vulnerable to bankruptcy.

INVESTMENT PROBLEMS AND STRATEGIES

Given the financial situation, U.S. integrated producers limited capital commitments and sought alternative sources of funding. Project financing, leveraged leasing, tax-oriented leasing, and joint ventures are among the financing approaches used.

Table A.12 shows U.S. steel industry capital expenditures from 1960 to 1986. Since 1983, total capital expenditures and total pollution control expenditures declined. They reached their lowest level in 1986. In 1982-83, the U.S. steel industry reduced capital expenditures from $2.3 billion to

Table A.6. Earnings per Share for the Six Largest U.S. Integrated Steel Producers and Nucor--1973 to 1986 (dollars)

Year	Armco	Bethlehem	Inland	LTV	National	USS	Nucor
1973	2.24	4.72	4.39	4.08	5.27	3.86	0.51
1974	4.43	8.03	7.96	8.02	9.44	8.20	0.82
1975	2.47	5.54	4.43	1.05	3.10	6.34	0.62
1976	2.53	3.82	5.20	2.74	4.53	5.00	0.68
1977	2.45	-1.68	4.23	-2.52	3.12	1.65	0.96
1978	4.29	5.15	7.61	1.07	5.85	2.83	1.95
1979	4.82	6.31	6.27	5.02	6.56	-3.37	3.15
1980	5.04	2.77	1.38	3.95	4.42	5.77	3.31
1981	4.97	4.83	2.69	7.97	4.59	12.07	2.51
1982	-5.41	-33.64	-5.60	-3.20	-24.75	-3.72	1.59
1983	-10.27	-3.94	-4.76	-3.72	-8.46	-11.99	1.98
1984	-4.54	-2.91	-1.90	-5.23	-0.09	3.53	3.16
1985	0.54	-4.34	-7.37	-9.00	-1.53	1.95	2.73
1986	-7.04	-3.37	0.27	-34.91	-4.54	-7.44	2.17

Source: *World Steel Dynamics*, "Steel Strategist #14," 1987.

$2.0 billion, which *World Steel Dynamics* estimates imply is equivalent to cutting capacity at an average rate of 2 percent per year.

World Steel Dynamics also estimates that at an investment outlay of $16 per ton of existing capacity, output falls 2 percent per year; outlays of $25 per ton suffice to maintain capacity; expenditures of $46 per ton would lead to a 3 percent per year expansion. These estimates are based on costs of $2 per existing ton to meet pollution requirements, $10 per existing ton for repair and maintenance, replacement costs of $300 per additional ton, and capacity expansion costs of $800 per additional ton. With U.S. steel industry capacity of 100 million tons, $1.6 billion must be spent annually to prevent capacity from contracting more than 2 percent. Sustaining capacity costs about $2.5 billion per year. Another $200 million is needed for pollution control expenses.

The return on investment of a green-field plant using the conventional steel-producing technology for steelmaking (i.e., coke-ovens, blast furnace, and BOF furnaces) appears insufficient (see Chapter 6), and U.S. steel producers do not expect to see new integrated facilities built in the United States.

However, investment to modernize the finishing capacity and add new finishing facilities has been attractive. Capital investment declined drastically during the 1982-86 period, and the majority of the capital investments were directed to the finishing facilities.

During these three years, most of the capital expenditures were directed toward continuous casting and ladle metallurgy facilities. In 1986, eleven continuous casters and thirteen ladle metallurgy facilities were installed. Also, expenditures in the finishing of flat-rolled products increased due to the rising demand for electro-galvanized sheet. During 1984-86, five electro-galvanizing lines were installed. Stress is on securing the most modern equipment using the world's best available technologies. U.S. steel

Financial Situation of U.S. Producers

Table A.7. Stock Prices for the Six Largest U.S. Integrated Steel Producers and Nucor--1973 to 1986 (dollars per share)

Year	Armco	Bethlehem	Inland	LTV	National	USS	Nucor
1973	14.40	29.30	30.30	9.70	35.70	21.50	2.90
1974	14.70	29.30	31.00	9.70	33.20	27.80	2.40
1975	18.70	34.50	39.40	13.30	38.10	39.90	2.60
1976	21.10	41.20	51.40	13.40	47.00	51.50	4.60
1977	17.90	27.90	41.10	8.70	36.40	37.80	5.20
1978	19.50	22.10	36.80	7.40	30.90	26.10	8.90
1979	22.40	22.40	35.40	8.50	31.40	21.90	16.70
1980	31.40	24.00	29.90	12.90	27.40	20.90	26.80
1981	33.00	24.70	28.70	20.30	25.90	29.90	33.50
1982	17.90	18.70	21.60	12.00	17.70	21.10	23.60
1983	18.50	23.20	30.10	15.40	26.90	25.80	39.40
1984	14.50	20.80	23.80	13.10	28.40	26.20	32.40
1985	9.20	16.70	22.50	8.50	27.50	27.20	42.10
1986	8.30	12.10	21.10	4.70	20.20	21.60	41.00

Source: *World Steel Dynamics,* "Steel Strategist #14," 1987.

industry managers are seeking partners to provide the necessary technology, capital, and human resources.

Among the financing methods used, project financing is an arrangement in which steel companies recruit outside investors for major projects, rather than relying on their own credit worthiness to secure financing. Although the steel-maker manages the project, the costs and benefits associated with the ownership accrue to the outside investors. Leveraged leasing involves securing investment capital from outsiders to purchase a major piece of steelmaking equipment. The steel producer avoids immediate capital outlays but is committed to lease payments. Thus, the need to generate equity capital is delayed.

Tax-oriented leasing allowed steel producers to sell off accelerated depreciation and investment-tax-credit benefits to potential lessors. Since the steel industry was running losses and had no tax liabilities, these tax benefits had no value to them. Under tax provisions passed in 1981 and repealed in 1986, profitable companies could buy and use the credits to reduce their taxes. As with leveraged leasing, the lessor actually owns the equipment or project. However, the tax advantage lowered the costs to the lessor, and this produced reductions in rents charged to the steel producers.

In 1983, the General Electric Credit Corporation (GECC) was heavily involved in such deals because of the tax benefits available from steel producers. USS and National Steel sold tax benefits to GECC to finance new coke oven projects. Inland Steel used the same route to finance the second phase of a continuous annealing line.

Table A.8. Price Earnings Ratio for the Six Largest U.S. Integrated Steel Producers and Nucor--1973 to 1986

Year	Armco	Bethlehem	Inland	LTV	National	USS	Nucor
1973	6.43	6.21	6.90	2.38	6.77	5.57	5.69
1974	3.32	3.65	3.89	1.21	3.52	3.39	2.93
1975	7.57	6.23	8.89	12.67	12.29	6.29	4.19
1976	8.34	10.79	9.88	4.89	10.38	10.30	6.76
1977	7.31	-16.61	9.72	-3.45	11.67	22.91	5.42
1978	4.55	4.29	4.84	6.92	5.28	9.22	4.56
1979	4.65	3.55	5.65	1.89	4.79	-6.50	5.30
1980	6.23	8.66	21.67	3.27	6.20	3.62	8.10
1981	6.64	5.11	10.67	2.55	5.64	2.48	13.35
1982	-3.31	-0.56	-3.86	-3.75	-0.72	-5.67	14.84
1983	-1.80	-5.89	-6.32	-4.14	-3.18	-2.15	19.90
1984	-3.19	-7.15	-12.53	-2.50	-315.56	7.42	10.25
1985	17.04	-3.85	-3.05	-0.94	-17.97	13.95	15.42
1986	-1.18	-3.59	78.15	-0.13	-4.45	-2.90	18.89

Source: *World Steel Dynamics*, "Steel Strategist #14," 1987

JOINT VENTURES

Joint ventures attracted domestic and international partners. Most of the joint ventures involved finishing-end activities, an area in which profitable investment opportunities were most apparent. Several domestic steel companies initiated joint ventures with other U.S. firms and such participation allowed diversification of product lines and facilitated the securing of marketing and technical knowledge and capital (U.S. International Trade Commission, 1988b).

USS committed itself to two domestic joint ventures, one of which is a 540,000-ton processing facility in conjunction with Worthington Steel to process wide sheets from USS for customers in the automotive, appliance, furniture, and metal-door industries. Worthington Steel will provide 50 percent of the starting capital and manage the operation. USS also has entered a joint venture with Rouge Steel, a subsidiary of Ford, to cooperate on a two-sided electro-galvanizing line near Detroit, Michigan. Also involved in a new joint venture electro-galvanizing project are Inland Steel, Bethlehem, and Pre-Finish Metals Inc. (U.S. International Trade Commission, 1988b).

Foreign involvement in the U.S. flat-rolled product market, including both joint ventures with U.S. integrated producers and total ownership of steel operations, increased since 1981. Joint ventures between U.S. steel producers and foreign investors include the following:

--the joint venture of Inland and Nippon Steel to add a $250 million cold-rolling mill;

Financial Situation of U.S. Producers

Table A.9. Moody's Bond Rating of the Six Largest U.S. Integrated Steel Producers--1980 to 1987 [a]

	February 1980	1981	1982	1983	As of January 1984	1985	1986	1987
Armco	A	A	A	A2	Baa2	Baa3	Ba2	B2
Bethlehem	A	A	A	Baa2	Baa2	Ba1	Ba1	B2
Inland	Aa	A	A	Baa2	Baa2	Baa2	Baa3	Ba3
LTV (b)	n.a.	n.a.	n.a.	n.a.	n.a.	B1	B1	Ca
J & L (b)	Ba	Ba	Ba	Ba1	Ba1	Ba1	B3	Baa
Republic [b]	A	A	A	Baa3	Ba1	Ba1	B3	Caa
National	Aa	A	A	Baa3	Ba1	Ba1	B3	Ba3
USS	Aa	A	A	A3	Baa2	Baa2	Baa2	Ba1

Note: [a] Moody's bond rating are as follows:
Aaa: Best Quality, carrying the smallest degree of risk.
Aa: High Quality, Ranked together with Aaa as high-grade bonds.
A: Possessing many favorable investment attributes and considered upper-medium grade obligations.
Baa: Medium-grade obligations, neither highly protected nor poorly secured.
Ba: Obligations that have speculative elements; their future cannot be considered well-assured.
B: Generally lacking characteristics of desirable investment.
Caa: In poor standing; may be in default or may present elements of danger with respect to principal or interest.
Ca: Speculative in a high degree.
C: Lowest-rated bonds.
In 1983, Moody's modified its ratings. The numbers place the bond's rating within the alphabetic rating. 1 is preferable to 2, which is preferred to 3.
[b] LTV merged Jones Laughlin and Republic Steel during 1984; therefore the ratings are not available for LTV.
Source: U.S. International Trade Commission, *U.S. Global Competitiveness: Steel Sheet and Strip Industry*, Report 2050, 1988.

--the National Steel ownership on a fifty-fifty percent basis with National Intergroup and Nippon Kokan (speculations prevail that the National Intergroup intends to sell its entire steel operation to Nippon Kokan in 1989);
--the LTV and Sumitomo partnership in a $130 million electro-galvanizing line;
--the Nisshin Steel and Wheeling-Pittsburgh joint venture of a zinc and aluminum-coating facility, and the role of Nisshin Steel in the operations of Wheeling-Pittsburgh Steel, which includes placement of two Nisshin executives on the board of directors;
--the USS joint venture with Posco of South Korea to finish South Korean semi-finished steel at the Pittsburgh plant; and,
--Kawasaki ownership of 40 percent of Armco Steel involving $350 million.

Table A.10. Indicators of the Financial Condition of the U.S. Steel Industry--1979 to 1987

Year	Profit Margin[a] (percent)	Current Ratio[b] (ratio)	Acid Test[c] (ratio)	Debt/ Capitalization[d] (percent)	Working Capital[e] (billion of dollars)
1979	2.00	1.68	0.21	37.95	4.22
1980	1.83	1.65	0.18	40.26	4.25
1981	3.84	1.66	0.17	38.58	4.43
1982	-11.98	1.49	0.18	50.06	2.61
1983	-9.12	1.33	0.11	57.46	1.89
1984	-0.10	1.41	0.07	52.93	2.32
1985	-6.49	1.27	0.04	61.10	1.49
1986	-16.88	1.59	0.10	84.38	2.40
1987	3.79	1.52	0.12	82.15	2.65

Note: [a] The ratio of net income to net sales revenue.
[b] The ratio of current assets to current liabilities.
[c] The ratio of monetary current assets to current liabilities.
[d] The ratio of non-current liabilities to non-current liabilities plus shareholders' equity.
[e] Current assets minus current liabilities.
Source: Compiled from American Iron and Steel Institute, *Annual Statistical Report*, 1987.

Totally foreign-owned investment in the U.S. flat-rolled product market includes:
--the ownership on a fifty-fifty percent basis of Kawasaki and CVRD (the Brazilian iron-ore company) of California Steel Industries (CSI, the operator of facilities formerly part of now bankrupt Kaiser Steel);
--the West German, Friedr Gustav Theis Kaltwaltzerke GMbH, ownership of Theis Precision Steel Corp. in Connecticut, which process hot-rolled carbon sheets into cold-rolled sheets;
--the West German, Hille and Mueller GMbH, ownership of Thomas Steel Strip Corp. in Ohio and Illinois, which consists of cold-rolling, electro-galvanizing, and other coating facilities of carbon sheet and strip; and,
--the Canadian, Dofasco, ownership of Whittar Steel Strip in Michigan, which only re-processes hot-rolled sheets into cold-rolled sheets.

Most foreign involvement in the U.S. flat-rolled market involves the supply of equipment and technical and managerial assistance. Foreign participation in the U.S. steel industry should increase because of the growing international nature of the steel industry, the import-retarding effects of trade barriers, and exchange rate hedging inspired by variability in exchange rates.

Table A.11. Debt as a Percent of Equity[a] for the Six Largest U.S. Integrated Steel Producers--1984 to 1986

Year	Armco	Bethlehem	Inland	LTV	National	USS
1984	70	110	70	160	60	110
1985	60	160	100	330	70	80
1986	130	180	70	6	70	100

Note: [a] In calculating the percents, long-term debt plus the value of leases was used.
[b] In 1986, LTV had negative equity.
Source: U.S. International Trade Commission, *U.S. Global Competitiveness: Steel Sheet and Strip Industry*, Report 2050, 1988.

RESTRUCTURING

Reorganizations, which were nonexistent before 1981, increased thereafter. The most noticeable was that of LTV Steel. Other companies that experienced reorganization were McLouth, Weirton, CF & I, Wheeling-Pittsburgh, California Steel Industries, Gulf States, Newport, and Phoenix. Among the nine, only two, Newport and Phoenix, were of non-integrated producers. In 1986, these nine steel producers accounted for 25.5 percent of domestic shipments (*World Steel Dynamics*, 1987).

In addition to transferring their underfunded pension plans to the Pension Benefit Guarantee Corporation (PBGC), companies were able to reduce costs by canceling high-cost raw material contracts, re-negotiating labor contracts, and suspending interest payments while in bankruptcy. These measures have lowered the companies' production costs by as much as 20 percent.

Companies that are in or that have emerged from bankruptcy have secured the cost reductions such reorganizations produce. Table A.13 compares 1987 production costs of reorganized producers to those of other integrated firms. The employment cost per ton shipped is $48 lower due to improved work rules and reduced pension costs. By canceling high-cost contracts for iron ore and production by $13 and $5 per ton shipped.

Re-negotiation of contracts for other inputs and services contributes to a reduction of $11 per ton shipped. It has been estimated that the companies operating under Chapter 11 reduce wage costs about $3 per hour. This gives them a significant cost advantage over the other major mills (*World Steel Dynamics*, 1987).

Additionally, lower local and state taxes, creditors' loss, and lower depreciation expenses (a non-cash cost) produced a total reduction of $26 per ton shipped. The total cost reduction amounts to a maximum of $103 per ton shipped. The fixed costs reduction of $44 per ton reflects the repudiation of obligations to investors that is intrinsic to bankruptcy. Poor financial performance also forced many other companies to sell assets to generate the cash necessary to meet current obligations.

Table A.12. Total U.S. Steel Industry Capital Expenditures and Outlays for Pollution Control--1960 to 1986 (million of dollars)

Year	Water	Air	Solid Waste	Total Pollution	Total Expenditures
1960-70 average	315.9	256.9	n.a.	572.8	10,228.6
1971-75 average	439.3	790.6	n.a.	1,229.9	9,317.4
1976	158.7	330.5	n.a.	489.2	3,261.3
1977	205.7	329.1	n.a.	534.8	2,859.9
1978	180.8	277.2	n.a.	458.0	2,449.2
1979	201.2	449.6	n.a.	650.8	2,465.2
1980	168.2	342.3	7.5	518.0	2,656.4
1981	119.4	369.8	6.6	495.8	2,372.2
1982	104.6	157.3	2.6	264.5	2,260.7
1983	55.6	84.6	2.7	142.9	1,855.8
1984	53.3	79.6	4.8	137.7	1,207.9
1985	66.4	53.8	1.6	136.2	1,865.8
1986	31.5	39.3	1.6	72.4	882.9

Source: U.S. International Trade Commission, *U.S. Global Competitiveness: Steel Sheet and Strip Industry*, Report 2050, 1988 and American Iron and Steel Institute, *Annual Statistical Report*, 1987.

In 1987, the liabilities attributable to shutdown of facilities were $1.6 billion for the U.S. steel industry (AISI, 1987). Recent integrated steel plant closure costs have averaged $75,000 per employee, of which roughly 90 percent were cash costs. The single largest costs related to steel facility closures are employee-related and represent 72 percent of the total (Chase Econometrics, 1987).

When a company goes bankrupt the federal government's PBGC generally assures that workers covered by the plan receive their benefits. In return, the PBGC is entitled to certain assets of the bankrupt firm. A firm can thus be relieved of the liabilities of laying off workers by declaring bankruptcy and may then continue to operate under Chapter 11 of the Bankruptcy Code. This policy essentially subsidizes the least efficient firms, since they are the ones most likely to go bankrupt.

The issue of underfunded pension liabilities became a major focus of the task force from the Economic Policy Committee (EPC), especially after the January 1987 transfer of the LTV Corporation's underfunded pension plan to the PBGC, the government agency responsible for insuring pension plans. At the time, administration officials were concerned that LTV's use of the bankruptcy process to lower its production costs might encourage other companies to follow this precedent. LTV's action added $2.3 billion to the PBGC's $4 billion deficit. Potential claims against the government by steel companies that had not yet filed for bankruptcy were estimated to total an additional $4 billion to $6 billion.

Major integrated producers considered mergers as a way to cut costs. In 1984, Jones & Laughlin Steel merged with Republic Steel forming the LTV group. Paine Webber estimated that the Republic Steel and Jones & Laughlin Steel merger reduced the companies' combined capital spending

Table A.13. Cost Difference Between Reorganized and Other U.S. Integrated Producers in 1987 (dollars per metric ton shipped)

	Reorganized Integrated	Other Integrated	Cost Reduction
Operating Rate (percent)	80	90	
Labor	158	110	-48
Iron Ore	66	53	-13
Metallurgical Coal	47	42	-5
Other Inputs and Service Contracts	162	151	-11
Taxes, Depreciation, Interest	42	16	-26
Total Cost	475	372	-103
Fixed Costs[a]	114	70	-44
Operating Costs	433	356	-77

Note: [a] Fixed costs are considered to be 33 percent of labor costs, 7 percent of material costs and 100 percent of financial costs (including taxes).

Source: *World Steel Dynamics*, "Steel Strategist #14," 1987.

requirements by $700 million and cut their annual operating costs by $450 million. However, as shown above, bankruptcy still developed.

USS announced plans to acquire the steel operations of National Intergroup. Other steel producers such as Bethlehem were reported to be looking for mates. Except for the LTV merger, other merger attempts were opposed by the Justice Department. This reflects the practice of enforcing the anti-merger provisions of the Clayton Act by opposing mergers in industries in which high levels of "concentration" (high market shares by the top producers) prevail. Concentration is deemed an indicator that oligopoly may exist.

A standard criticism of use of concentration ratios is that they ignore foreign competition. Such competition would be formidable if imports were not controlled. Moreover, domestic price increases that substantially raise the cost of import controls tend to stimulate pressures to reduce protection. Thus, it is not clear whether domestic producers are so greatly shielded from import competition that they can behave oligopolistically. Traditional economic analysis suggests that removal of import restrictions would be preferable.

APPENDIX B

Closures of Steel Capacity

First, this appendix lists details of U.S. plant and facility closures. Then, it reviews the differences between crude-steel integrated and flat-rolled capacities reduction in the East and the West.

PLANT CLOSURES

Closures of wire-rod and bar capacity by integrated producers included partial or total deactivation at:

- --Armco's Houston and Torrance facilities;
- --Bethlehem's Los Angeles, Bethlehem, South San Francisco, Johnstown, Lackawanna, and Sparrows Point facilities;
- --Cyclops' Portsmouth facility;
- --LTV's Aliquippa, Campbell, Indiana Harbor, Warren, Buffalo, Gadsen (now Gulf States), and Youngstown facilities;
- --Ford's Rouge facility; and,
- --USS's Clairton, Cleveland, Duquesne, Duluth, Fairfield, Fairless, Gary, Joliet, Pittsburgh, South Chicago, Torrance, and Youngstown facilities (*World Steel Dynamics*, 1987).

The integrated producers also reduced flat-rolled capacity, including the partial or total closures of:

- --Armco's Houston;
- --Bethlehem's Lackawanna;
- --LTV's Pittsburgh plant; and
- --USS's Baytown (1987), Homestead (1987), Gary, South Chicago, and Youngstown plants (*World Steel Dynamics*, 1987).

Closures of Steel Capacity

These closures have produced a permanent closure of the following plants:

- --Armco's Torrance (1985) and Houston (1983) plants;
- --Bethlehem's Los Angeles (1982) plants;
- --Cyclops' Portsmouth (1980) plant;
- --LTV's Aliquippa (1985), Buffalo (1983), Brier Hill (1977), Campbell (1979), Pittsburgh (1987), South Chicago (1987), and Youngstown (1981) plants;
- --USS's Cleveland (1984), Duquesne (1987), Torrance (1979), Youngstown (1979), and Homestead (1983) plants;
- --Wheeling-Pittsburgh's Monessen (1987) plant; and
- --Wisconsin Steel (Chase Econometrics, 1987, and *World Steel Dynamics*, 1987).

Others included partial closures of coking, blast furnaces, and OH and BOF furnaces. In some cases, the plants reduced or eliminated their vertical integration. Lone Star Steel (1987) and Colorado Fuel and Iron (1984) became non-integrated producers. Weirton and Sharon depend on external sources of coke. Some plants belonging to Armco, Bethlehem, LTV, and USS are shipping intermediate products to be processed further in another plant of the same company. One example of this is the re-processing of billets from Bethlehem Steel's Johnstown plant at the company's Lackawanna plant, which completely closed its hot-end facilities.

Some integrated producers making non-flat-rolled products have switched steel production to electric furnaces or increased their electric furnace capacity to improve their ability to compete with mini-mills. This is the case with LTV's Pittsburgh plant, Bethlehem's Johnstown plant, and Armco's Kansas City plant. Even some integrated producers making flat-rolled products have increased EF production. This is true for USS's Baytown plant, LTV's Cleveland and Midland plants, and National's Great Lakes plant.

THE REGIONAL DISTRIBUTION OF CRUDE STEEL INTEGRATED CAPACITY

As shown above, U.S. crude-steel capacity declined from 1977 to 1986. This decline was proportionally greater in the West than in the East. Table B.1 presents the regional distribution of U.S. crude-steel capacity and the characteristics of the two regions.

U.S. crude-steel capacity is concentrated in the East, especially in the Northeast region. From 1977 to 1986, Eastern crude-steel capacity declined from 134 million to 112 million tons; that in the West, from 11 million to 6 million tons. Thus, Eastern capacity dropped 17 percent; Western, 47 percent.

Eastern crude-steel production fell from 106 million to 71 million tons, resulting in an operating rate decline from 79 percent to 63 percent. Western

Table B.1. Regional Distribution of the U.S. Crude Steel Capacity--1977 to 1986

		1977	1979	1980	1983	1986
East						
Capacity	(million metric tons)	134.3	130.7	129.8	129.0	112.2
Integrated	(percent)	89.1	87.9	86.4	83.0	79.2
Non-integrated	(percent)	10.9	11.1	13.6	17.0	20.8
BOF	(percent)	60.5	63.6	63.6	60.9	59.6
EF	(percent)	21.6	24.2	26.3	30.5	35.5
OH	(percent)	17.9	12.2	10.1	8.6	4.9
Production	(million metric tons)	106.4	115.7	94.8	72.8	70.8
Op. Rate	(percent)	79.3	88.5	73.0	56.5	63.1
West						
Capacity	(million metric tons)	11.0	11.5	10.7	8.7	5.9
Integrated	(percent)	67.8	71.8	64.7	50.2	30.7
Non-integrated	(percent)	32.2	28.2	35.3	49.2	69.3
BOF	(percent)	21.8	38.3	34.1	29.2	0
EF	(percent)	41.2	36.8	44.4	57.0	79.6
OH	(percent)	37.0	24.9	21.5	13.8	20.4
Production	(million metric tons)	7.2	8.0	6.7	3.9	3.3
Op. Rate	(percent)	65.9	69.8	63.0	44.6	55.3

Source: Compiled from *World Steel Dynamics*, "Core Report BB," 1988 and American Iron and Steel Institute, *Annual Statistical Report,* Various Years.

crude-steel production declined from 7 million to 3 million tons, and operating rates went from 66 percent to 55 percent. In the East, integrated producers' share of capacity declined from 89 percent to 79 percent, and in the West, from 68 percent to 31 percent.

At least 19 percent of the decline of the integrated share in the West resulted from the shutdown of the Kaiser plant toward the end of 1983. The plant was purchased by Kawasaki and Rio Doce Ltd. for $110 million but only the finishing portions still operate. In 1986, the only integrated plant operating was USS's Geneva Works. This plant was closed in mid-1986 during a company-wide work stoppage and did not reopen after the strike. Since its capacity was included in the 1986 figure, the shutdown was another contributor to a lower operating rate.

In 1987, Basic Manufacturing and Technology, Inc. of Salt Lake City purchased the plant and restarted independent operation as Geneva Steel. Geneva crude-steel production, based on OH furnaces, has no continuous-casting facilities and without modernization, its survival as an integrated producer is uncertain.

Thus, the West has no slab continuous-casting facilities, and its only integrated steel producer uses OH furnaces. Therefore, the flat-rolled production of the region had to rely on slab purchases for quality products from the East and from imported suppliers.

Table B.2. Regional Distribution of the U.S. Flat-Rolled Capacity--1977 to 1986

		1977	1979	1980	1983	1986
East						
Reversing Mill	(million metric tons)	11.1	11.2	10.8	10.6	8.2
Integrated	(percent)	85.7	85.7	85.3	85.0	80.5
Non-Integrated	(percent)	14.3	14.3	14.7	15.0	19.5
Hot-Strip Mill	(million metric tons)	72.5	72.4	69.1	67.1	64.9
Integrated	(percent)	99.2	99.1	99.1	99.1	98.6
Non-Integrated	(percent)	0.8	0.9	0.9	0.9	1.4
Total Flat-Rolled	(million metric tons)	83.6	83.6	79.9	77.7	73.0
Integrated	(percent)	97.4	97.4	97.2	97.2	96.6
Non-Integrated	(percent)	2.6	2.6	2.8	2.8	3.4
West						
Reversing Mill	(million metric tons)	1.2	1.2	1.2	1.2	0.4
Integrated	(percent)	64.1	64.1	64.1	64.1	0.0
Non-Integrated	(percent)	35.9	35.9	35.9	35.9	100.0
Hot-Strip Mill	(million metric tons)	3.5	3.5	3.6	3.8	3.5
Integrated	(percent)	93.4	93.4	93.6	94.0	53.4
Non-Integrated	(percent)	6.6	6.6	6.4	6.0	46.6
Total Flat-Rolled	(million metric tons)	4.6	4.6	4.8	5.0	3.9
Integrated	(percent)	86.0	86.0	86.4	86.9	47.7
Non-Integrated	(percent)	14.0	14.0	13.6	13.1	52.3

Source: Compiled from *World Steel Dynamics*, "Capacity Monitor # 2," 1987.

TRENDS IN FLAT ROLLED CAPACITY

In 1984-86, the share of other producers in the flat-rolled capacity increased because of CSI, and to a lesser extent Tuscaloosa (Alabama). The CSI plant started its operations in 1984; Tuscaloosa, in 1986. Both purchase slabs that are transformed into flat-rolled products.

Table B.2 shows the distribution between Eastern and Western flat-rolled capacity from 1977 to 1986. The non-integrated share of U.S. reversing mill capacity grew from 13 percent to 19 percent. The share of hot-strip mill capacity for non-integrated producers only increased from 1 percent to 4 percent. Eastern flat-rolled capacity fell from 84 million to 73 million tons, but integrated producers maintained their share of total Eastern capacity. However, integrated producers' share of reversing mill capacity declined from 86 percent to 81 percent. Conversely, the share of the hot-strip mill remained almost unchanged since the influence of the Tuscaloosa plant was very small.

Between 1977 and 1986, Western flat-rolled capacity declined from 5 million to 4 million tons. This decline resulted from a decrease in reversing mill capacity; hot-strip mill capacity was maintained at 4 million tons. The integrated producers' share of Western capacity, however, fell for reversing and hot-strip mills. The most noticeable change was for reversing mills. Integrated-firm capacity was formerly 64 percent of the total; no integrated-firm capacity remains. For the hot-strip mill capacity, integrated producers share fell from 93 percent to 53 percent. This resulted in integrated share of the total flat-rolled capacity dropping from 86 percent to 48 percent.

In 1984, non-integrated producers increased their inroads into flat-rolled capacity. Much of this growth of the share of the U.S. flat-rolled capacity was in the West. To a large extent, non-integrated penetration in the flat-rolled market was concentrated in the plate market (using reversing mills) because plates are produced in EF furnaces, and scrap contamination is not a limitation. This is not true for sheet production because most sheet applications require a low concentration of impurities.

Thus, the restructuring of the industry most severely affected the Western region, in which crude-steel capacity declined by 47 percent from 1977 to 1986, and no continuous casting capacity remains. Therefore, most of the flat-rolled producers have become dependent on slab purchases, especially for quality sheet production.

APPENDIX C

Capacity Balance Equations

This appendix introduces the capacity balance equations used for producing Tables 5.10 to 5.13. The capacity data used were provided by *World Steel Dynamics*.

CAPACITY BALANCE EQUATION FOR COKE OVEN AND BLAST FURNACE CAPACITIES IMBALANCE

This equation was used for Table 5.12 for showing the imbalance between coke ovens and blast furnace capacities:

$$\sum_{x=1}^{n} [OP_{CK}\, CK_{xy} - OP_{BF}\, [CR \cdot BF_{xy}]] = CI_y$$

where: CI = excess capacity (tons of coke), where
$CI > 0$ means surplus of coking capacity, and
$CI < 0$ means lack of coking capacity.
OP_{CK} = coke oven operating rate (percent)
CK = coke oven capacity (million metric tons)
OP_{BF} = blast furnace operating rate (percent)
BF = blast furnace capacity (million metric tons)
CR = coke rate (ton of coke/ ton of pig iron)
x = plant
y = year

CAPACITY BALANCE EQUATION FOR BLAST FURNACE AND STEELMAKING CAPACITIES IMBALANCE

This equation was used for Table 5.13 for showing the imbalance between blast furnace and steelmaking capacities:

$$\sum_{x=1}^{n}[OP_{BF}BF_{xy}-[OP_{BOF}BOF_{xy}(1-SC)+OP_{OH}\ OH_{xy}\ (1-SC)]] = PI_y$$

where: PI= excess of blast furnace capacity (tons of pig iron), where
PI > 0 means surplus of blast furnace capacity, and
PI < 0 means lack of blast furnace capacity.
OP_{BF}= blast furnace operating rate (percent)
BF= blast furnace capacity (million metric tons)
OP_{BOF}= basic oxygen furnace operating rate (percent)
BOF= basic oxygen furnace capacity (million metric tons)
OP_{OH}= open-hearth furnace operating rate (percent)
OH= open-hearth furnace capacity (million metric tons)
SC= percent of scrap charged (percent)
x= plant
y= year

CAPACITY BALANCE EQUATION FOR STEELMAKING AND FLAT-ROLLED FINISHING CAPACITIES IMBALANCE

This equation was used for Tables 5.10 and 5.11 for showing the imbalance between steelmaking and flat-rolled finishing capacities.

$$\sum_{n}^{x=1}[[OP_{BOF}BOF_{xy}+OP_{OH}OH_{xy}+OP_{EF}EF_{xy}][PS(1-CC)Y_{IC}+PSCC\ Y_{CC}]]$$

$$-\sum_{x=1}^{n}\left[OP_{PM}\cdot\frac{PM_{xy}}{Y_{PM}}+OP_{HS}\cdot\frac{HS_{xy}}{Y_{HS}}\right]=SI_y$$

where: SI = excess steelmaking capacity (tons of slab), where
SI > 0 means surplus of steelmaking capacity, and
SI < 0 means lack of steelmaking capacity.
OP_{BOF}= basic oxygen furnace operating rate (percent)
BOF= basic oxygen furnace capacity (million metric tons)
OP_{OH}= open-hearth furnace operating rate (percent)
OH= open-hearth furnace capacity (million metric tons)
OP_{EF}= electric furnace operating rate (percent)
EF= electric furnace capacity (million metric tons)
PS= liquid steel directed for flat-rolled products production, where PS is equal to the company product mix when total steelmaking capacity of all plants were used, and is equal to the share of flat-rolled capacity from total finishing capacity when only plants producing flat-rolled products were considered. In this last case, only the BOF, OH and EF capacities of plants producing flat-rolled products were considered.

Capacity Balance Equations

CC= share of continuous casting (percent)
Y_{IC}= ingot casting yield (ton of slab/ ton of liquid steel)
Y_{CC}= yield of continuous casting (ton of slab/ ton of liquid steel)
OP_{PM}= plate mill operating rate (percent)
PM= plate mill capacity (million metric tons)
Y_{PM}= plate mill yield (ton of plate/ ton of slab)
OP_{HS}= hot-strip mill operating rate (percent)
HS= hot-strip mill capacity (million metric tons)
Y_{HS}= hot-strip mill yield (ton of hot-rolled coil/ ton of slab)
x= plant
y= year

APPENDIX D

Cost Estimation Procedure

I use a simple methodology for estimating costs from information on the use and unit cost of inputs. My system begins by identifying each cost component. For each, cost per ton of output is necessarily the price per unit of input times the number of inputs per ton. This is a standard procedure in costing. What is special here is that my research data, much of which are confidential, were obtained on the most appropriate price and unit input assumptions. The data were collected in interviews with steel consultants and steel producers and through the examination of specialized steel publications.

The cost procedures initially estimate the cost of producing coke and sinter, both intermediate products, by using the equation:

$$\sum_{i=1}^{m} \left(P_i \frac{X_i}{X_o} + \frac{M_i}{X_o} \right) = C_r$$

where: C_r = cost of coke or sinter
P_i = price of input
X_i = quantity of input
X_o = quantity of output
M_i = expenditures on miscellaneous items, for example, supplies and maintenance

The inputs used for the cost estimate of coke are coal, labor, electricity, other energies, and by-products, which are counted as a credit because they are consumed in other processes of steelmaking or sold. The inputs used for the cost estimate of sinter are sinter feed, coke fines, labor, electricity, other energy, and fluxes.

The equation used for calculating the cost estimates of hot metal (or pig iron) and of liquid steel is:

Cost Estimation Procedure

$$\sum_{i=1}^{m}\left(P_i \frac{X_i}{X_o} + \frac{M_i}{X_o}\right) + \sum_{r=1}^{n} C_r \frac{X_r}{X_o} = C_l$$

where: C_l = cost of hot metal or liquid steel
P_i = price of input
X_i = quantity of input
C_r = cost of coke, or sinter for the hot metal case and the cost of pig iron for the liquid steel case
X_r = quantity of coke, or sinter for the hot metal case and the quantity of pig iron for the liquid steel case
X_o = quantity of output
M_i = expenditures of miscellaneous items, for example supplies and maintenance

The inputs used for the estimates of producing hot metal are pellets, iron ore, labor, electricity, other energy, and by-product generation, the last of which, as in the case of coke ovens, is treated as a credit. For estimating the cost of producing liquid steel, the inputs used are scrap, ferro-alloys, aluminum, fluxes, oxygen, electricity, other energy, and by-products. Xr and Cr relate to different inputs depending upon whether hot metal or liquid steel costs are being considered. In the former case, the inputs are coke, sinter, and pellets in the latter hot metal. In estimating the cost of producing hot metal and liquid steel, the miscellaneous costs include reserves for relining blast furnaces for hot metal and the expenses of refractories for liquid steel.

As observed in Chapter 2, slabs can be produced either by continuous casting or by ingot casting. As a result, the cost estimate for slab production uses this equation:

$$CC\left[C_l \frac{X_l}{X_o} + \sum_{i=1}^{m}\left[P_i \frac{x_i}{x_o} + \frac{M_i}{X_o}\right]\right] + (1-CC)\left[C_l \frac{X_l}{X_o} + \sum_{i=1}^{m}\left[P_i \frac{x_i}{x_o} + \frac{M_i}{X_o}\right]\right] = C_s$$

where: C_s = cost of slab
CC = percent of continuous casting
P_i = price of input
X_i = quantity of input , which differs between continuous casting and other approaches
C_l = cost of liquid steel
X_l = quantity of liquid steel
X_o = quantity of output
M_i = expenditures of miscellaneous, for example, supplies and maintenance

The first part of the equation is the average cost of producing slab from continuous casting, and the second part is the average cost of producing slab from ingot casting. The inputs for continuous casting are labor, electricity, refractories, and scrap. The inputs for ingots casting are labor, electricity, other energy, and scrap. Scrap is always treated as credit because it is either used as an input for liquid steel production or sold.

The above cost estimate for slab is the average operating cost. To obtain the average total cost of producing slab, the depreciation and interest costs of the equipment used in producing slab are added. The sales and overhead costs, which consist of labor used in sales, shipment, and miscellaneous costs, are added considering that the plant will only produce slabs.

The general equation used for estimating the cost of any flat-rolled product is the equation:

$$\sum_{i=1}^{m}\left[P_i \frac{X_i}{X_0} + \frac{M_i}{X_0}\right] = C_f$$

where: C_f = cost of finished product
P_i = price of slab
X_i = quantity of input
X_0 = quantity of output
M_i = expenditures of miscellaneous items, for example supplies and maintenance

The inputs used in producing flat-rolled products are slab, labor, electricity, other energy, and scrap, the last of which is a credit for reasons discussed above. In the case of galvanized sheet, zinc also is used as an input. The expenditures on miscellaneous items also include the spending on rolls. For integrated producers, the price of slab is replaced by the estimated average cost of producing slab.

To obtain the total cost of producing flat-rolled products, the depreciation and interest costs of the equipment used in its production are added. Sales and overhead costs also are added.

Bibliography

Acs, Zoltan J., 1984, *The Changing Structure of the U.S. Economy: Lessons from the Steel Industry,* New York: Praeger Publishers.
_____, 1985, "Why Bankers Aren't Returning Mini-Steel's Phone Calls," *33 Metal Producing,* April, p. 62.
Aglietta, M., 1988, "Os Grandes Problemas da Moeda Internacional," *Revista de Economia Politica,* Vol.8, #2, April-June.
American Iron and Steel Institute, 1985, "Over 600 Steel Facilities Shut Down Since 1974," *Steel News,* November 18.
_____, 1986, "Assuring Steel's Competitiveness for the Automotive Industry," *Steel Comments,* February 28.
_____, 1988, "Direct Steelmaking: Europe, Japan, and the U.S. Program," *Steel Comment,* May 17.
_____, Various Years, *Annual Statistical Report,* Washington: American Iron and Steel Institute.
American Metal Market, 1983a, "USW Starts Advertising Campaign Against W-P's Brazil Slab Import Plan," August 10.
_____, 1983b, "USW in Anti-Import Campaign, and W-P Slates Monessen Recall," August 11.
_____, 1984, "Semi-Finished Steel: Import Lid Creates Supply Squeeze," November 21.
_____, 1988a, "Claims of Prohibitive Slab Prices May Bring USX Onto Capitol Hill," March 10.
_____, 1988b, "Quotas Said Not Responsible for Steel Price Hikes," March 28.
_____, 1988c, "USX's Slab Delivery Hit by Tuscaloosa," April 15.
_____, 1988d, "VRAs, Technology, Reinvestment Vie as Steel Strategies," June 3.
_____, 1988e, "Short-Supply Program's Misuse by U.S. Firms Possible, Exec Says," June 17.
Barnett, Donald F., 1985, "A Industria Siderurgica em Situacao Aflitiva: Estrategias Novas para uma Nova Era," *XIV Congresso Brasileiro de Siderurgia,* Instituto Brasileiro de Siderurgia.
_____, 1986, "Steel: Structural Change and Strategic Issues," *Steel Survival Strategies II,* June 25.

_____, 1988, "U.S. Profits, Capabilities and Resources," *Steel Survival Strategies III,* June 21.
Barnett, Donald F. and Robert Crandall, 1986, *Up from the Ashes--The Rise of the Steel Minimill in the United States,* Washington: Brookings Institution.
Barringer, W., 1986,"Steps Required to Obtain Additional Quota for Slab Imports," *Memorandum to Companhia Siderurgica Tubarao,* Wilkie Farr & Gallagher, March 4.
Baumol, William J. and Wallace E. Oates, 1988, *The Theory of Environmental Policy,* second edition, Cambridge: Cambridge University Press.
Bright, S.L., 1984, "The Economics of the Steel Trigger Price Mechanism," *Business Economics.*
Burnier, Paulo, 1988, "Global Steel Industry--What's Next?," *Steel Survival Strategies III,* June 22.
Business Week, 1983, "Time Runs Out for Steel," June 13, p. 84.
_____, 1984, "For Steelmakers, No Mergers May Mean More Bankruptcies," March 5, p. 76.
_____, 1986, "One OPEC Meeting Won't Get U.S. Drilling Back Up to Speed," August 25, p. 62.
Calarco, Vincent J., 1988, *The Coal Situation,* Vol. 8, #3, July, New York: Chase Manhattan Bank.
Cantor, D.J., 1988, *Steel Prices and Import Restraints,* Washington: U.S. Congressional Research Service.
Chase Econometrics, 1987, "New Uncertainties in the U.S. Steel Market."
Corrigan, E.G., 1988, "Um Instantaneo da Economia Mundial," *Economic Impact,* #60, May 12.
Crandall, Robert W., 1981, *The U.S. Steel Industry in Recurrent Crisis,* Washington: Brookings Institution.
Eggert, Roderick G., 1986, "Changing Patterns of Materials Use in the U.S. Automobile Industry," Paper Presented at the Metals Demand Conference, The Pennsylvania State University.
Financial Times, 1988, "Kawasaki Buys 40% of Armco Core Steel Business for $350 M," November 23.
Franz, J., 1986, "Iron Ore, Global Prospects for the Industry, 1985-95," *Industry and Finance Series,* Vol. 12, Washington: The World Bank.
Friden, L., 1972, *Instability in the International Steel Market,* London: Beckmans Publishers.
Fritsch, Wolfang, 1988, "Perspectivas da Economia Internacional," *RBCE,* #15, January-February.
Goldberg, Walter, 1986, *Ailing Steel--The Transoceanic Quarrel,* New York: St. Martin's Press.
Gordon, Richard L., 1987, *World Coal,* Cambridge: Cambridge University Press.
Graham, T.C., 1986, *Minerals Trade and Commercial Policy: The Case of Steel,* U.S.S. Division, U.S.X. Corp.
Hageman, R.A., 1985, "Steel--Yesterday and Today," *I&SM,* February, p. 35.

Bibliography

Hogan, William T., 1984, *Steel in the United States: Restructuring to Compete*, Lexington: D.C. Lexington Books.

Innace, J.J., 1983, "Easing Steel's Cash Squeeze," *33 Metal Producing*, May, p. 67.

_____, 1985a, "DRI in the Developed World," *33 Metal Producing*, October, p. 35.

_____, 1985b, "The Long, Quiet Twilight of the American Blast Furnace," *33 Metal Producing*, November, p. 41.

_____, 1986, "Bending the Rules in Basic Oxygen Steelmaking," *33 Metal Producing*, April, p. 39.

Instituto Brasileiro de Siderurgia, 1985, *Anuario Estatistico da Industria Siderurgica Brasileira*, IBS Yearbook, Instituto Brasileiro de Siderurgia.

International Energy Agency, 1988a, *Energy Policies and Programmes of IEA Countries*, Paris: Organization for Economic Cooperation and Development.

_____, 1988b, *Coal Information 1988*, Paris: Organization for Economic Cooperation and Development.

International Iron and Steel Institute, Committee of Economic Studies, 1980, "Causes of the Mid-1970's Recession in Steel Demand," Brussels: International Iron and Steel Institute.

International Iron and Steel Institute, 1987a, "Intermaterial Competition--An Economic Analysis of General Trends," Brussels: International Iron and Steel Institute.

_____, 1987b, *Steel Statistics Yearbook*, Brussels: International Iron and Steel Institute.

International Monetary Fund, Various Years, *International Financial Statistics*.

Iron Age, 1984, "Older Mills Are Competitive Today," November 5, p. 51.

Iron and Steel Statistics Bureau, Various Years, *World Steel Statistics Summary*, London: Iron and Steel Statistics Bureau.

Isenberg-O'Laughlin, Jo, 1985, "Thin-Slab Casting: Has Its Time Come?" *33 Metal Producing*, March, p. 43.

Jacobson, John E., 1986, "Foreign Involvement in the U.S. Steel Industry," *MBM*, May, p. 9.

Kim, J., 1986, The Korean Steel Industry, M.S. Thesis in Mineral Economics, The Pennsylvania State University.

Korf Engineering GmBH, n.d.,"KR Coal Reduction--Corex," Company Prospects of Technology.

Kountze, M. J., 1985, "Why the Oil Glut Is Good News to America's Mills," *33 Metal Producing*, March, p. 56.

Kutscher, H., 1986, "Forces for Restructuring the Industry and Its Facilities," *Proceedings of the Conference Sponsored and Organized by the Ironmaking and Steelmaking Committee and Engineering Committee of the Institute of Metals*, The Institute of Metals, May.

Larson, Edward D., 1986, "Beyond the Era of Materials," *Scientific American*, Vol. 254, #6, June, p. 24.

McManus, George, 1984a, "Steel Recast Itself," *Iron Age*, January 4, p. 81.

_____, 1984b, "A New Generation of Hot Strip Mills," *Iron Age*, May 21, p. 44.

_____, 1985, "Electric Furnace Report: Growth That Won't Stop,"*Iron Age*, January 2, p. 107.

_____, 1986, "GM Puts Pressure on Hot Strip Mills," *Iron Age*, January 3, p. 53.

Metal Bulletin, 1984a, "Semis Concept Still Alive," March 9, p. 7.

_____, 1984b, "Continuous Casting World Reference List," September, p. 93.

_____, 1986, *Metal Bulletin Handbook, Iron and Steel Statistics*.

_____, 1987, "Tightness in Slabs Boost Export Prices," October 26, p. 23.

_____, 1988a, "Steel Needs to Invest Report Warns, and Iscor's Start-Up of Corex Plant Delayed," March 30, p. 22.

_____, 1988b, "Russian Backing for New Brazilian Project," March 31, p. 19.

_____, 1988c, "New Slab Mini Plan in USA," March 31, p 22.

_____, 1988d, "Congress Acts to Relieve Slab Shortage," April 11, p. 31.

_____, 1988e, "Kobe in New Slabs for Export Project," April 14, p. 23.

_____, 1988f, "Tuscaloosa Latest to Seek More Slabs," April 18, p. 19.

33 Metal Producing, 1985b, "World Steel Industry Data Handbook--USA."

_____, 1986, "Darlington Revisited: A Thin-Slab Casting Update," April, p. 49.

_____, 1987, "New Directions in Rolling and Finishing," September, p. 25.

Miller, Joseph R., 1984, "Steel Minimills," *Scientific American*, Vol. 250, No. 5, May, p. 33.

Mueller, Hans, 1984, *Protection and Competition in the U.S. Steel Market*, Monograph Series #30, Murfreesboro: Tennessee State University.

_____, 1988a ,"Protection and Market Power in the Steel Industry," *Challenge*, September-October.

_____, 1988b, "Comercio de Aco em Um Contexto de Taxas Cambiais Insconstantes, de Intervencao e Mudancas Technologicas," *Servico de Divulgacao e Relacoes Culturais dos Estados Unidos da America,*

_____, 1988c, "Uma Visao do Comercio Mundial de Aco," *Congresso de Comercio Exterior*.

News from the Steelworkers, 1983, "USWA Will Fight Plan to Import Brazilian Slabs," United Steelworkers of America, July 27.

New York Times, 1984, "A Restructured Steel Industry," Feb. 2, p. 1.

Pfeffermann, G., 1985, "A Supervalorizacao das Taxas de Cambio e o Desenvolvimento," *Financas e Desenvolvimento*, p. 17, March.

Scheuerman, William, 1986, *The Steel Crisis--The Economics and Politics of a Declining Industry*, New York: Praeger Publishers.

Bibliography 181

Smith, Adam, 1776, *An Inquiry into the Nature and Causes of the Wealth of Nations*, Glasgow Edition, Oxford: Oxford University Press and Indianapolis: Liberty Classics, 1979.

Soares, Rinaldo C., 1987, "Perspectivas Tecnologicas Para A Industria Siderurgica Brasileira," *Metalurgia ABM*, Vol. 43, # 357, August, p. 40.

Steel Times, 1986, "Restructuring Steel Plants for the Nineties," July, p. 352.

Stigler, George J., 1965, *Essays in the History of Economics*, Chicago: University of Chicago Press.

_____, 1968," The Division of Labor is Limited by the Extent of the Market," *The Organization of Industry*, Homewood: Richard D. Irwin.

Szekely, Julian, 1978, *The Future of the World's Steel Industry*, Proceedings of the Fifth C.C. Furnas Memorial Conference, New York: Marcel Dekker Inc.

Tarr, D.G. and M.E. Morkre, 1984, *Aggregate Costs to the United States of Tariffs and Quotas on Imports: General Tariff Cuts and Removal of Quotas on Automobiles, Steel, Sugar, and Textiles*, Washington: U.S. Federal Trade Commission.

Teplitz, B., 1984, "Third World Steel Gets Here Directly or Indirectly," *American Metal Market*, July 17, p. 10.

Tex Report, 1988, *1988 Coal Manual*, Tex Report Limited.

Tony, W. A., 1988, "Is a Coke Shortage Around the Corner?," *I&SM*, March, p. 16.

U.S. Congressional Budget Office, 1984, "The Effects of Import Quotas on the Steel Industry," Washington: U.S. Government Printing Office.

_____, 1986, "How Federal Policies Affect the Steel Industry," Washington: U.S. Government Printing Office.

U.S. Department of Commerce, 1986, *U.S. Industrial Outlook*.

_____, Various Years, *Current Industrial Reports--Steel Mill Products*, Bureau of Census, MA33B.

U.S. International Trade Commission, 1988a, *Annual Survey Concerning Competitive Conditions in the Steel Industry and Industry Efforts to Adjust and Modernize*, Publication 2115, Washington: U.S. International Trade Commission.

_____, 1988b, *U.S. Global Competitiveness: Steel Sheet and Strip Industry*, Publication 2050, Washington: U.S. International Trade Commission.

United States Steel, 1985, *The Making, Shaping, and Treating of Steel*, 10th Edition, Association of Iron and Steel Engineers.

Wall Street Journal, 1985, "Big Steel Is Facing Major Structural Changes," August 19, p. 6.

_____, 1986, "U.S. Steel, Union Are at Odds on Eve of Talks," June 11, p. 6.

_____, 1988, "Uneasy Revival: Big Steel Is Making a Comeback, But the Upturn Has Been Costly and May Not Last," November 4, p. A7.

World Steel Dynamics, 1986a, "WSD Capacity Monitor #2," Report #22.

_____, 1986b, "Steel Strategist #13," New York: Paine Webber.

_____, 1987a, *Global Steelmaking Capacity Track*, Core Report BB, New York: Paine Webber.

_____, 1987b, "Recurring Surprises in the Domestic and International Steel Market," Steel Survival Strategies II, June 24.

_____, 1987c, "Steel Strategist #14," New York: Paine Webber.

_____, 1988a, "The Steel Outlook: Key Perspectives and Plausible Wild Cards, " Steel Survival Strategies III, June 21.

_____, 1988b, " Cost Monitor," New York: Paine Webber.